Jewels of the Early Earth

Minerals and Fossils of the Precambrian

Bruce L. Stinchcomb

4880 Lower Valley Road, Atglen, Pennsylvania 19310

Acknowledgments

Many persons contributed to this work. The author would like to specifically acknowledge the following persons for recent assistance involving photography of specimens in their collections. Richard Hagar of Barnhart Missouri, Joseph A Lobacz, Jr. of St. Louis, Curvin Metzler of Alaska and Tucson, Arizona, Larry Nuelle of Brass Rooster Minerals, Rolla, Missouri, and Glenn Williams of St. Louis. I also wish to thank all of them for informative dialog. Through associations with the geology department at Florissant Valley Community College, where I taught, thanks are given to the following persons who, with the assist and burden of a bulky 35mm camera, captured excellent photos in the field. I am pleased to exhibit some of these, which are the work of Warren Wagner and Pete Kellams. I should also like to acknowledge the following students at Florissant Valley Community College Geology Dept., who offered companionship and stimulation on long trips in the "Blue Bomber" to far flung places like Labrador and central Canada. Martin Connelly, Jim Kocher, Dave Rickman, and the late Gene Mohr. Also I would like to thank Jeff Snyder for his many helpful assists, especially with place names.

Other Schiffer Books By The Author:

Cenozoic Fossils I: Paleogene. ISBN: 9780764334245. $29.99
Cenozoic Fossils II: The Neogene. ISBN: 9780764335808. $29.99
Mesozoic Fossils I: Triassic and Jurassic. ISBN: 9780764331633. $29.99
Mesozoic Fossils II: The Cretaceous Period. ISBN: 9780764332593. $29.99
Meteorites. ISBN: 9780764337284. $29.99
Paleozoic Fossils. ISBN: 9780764329173. $29.95
World's Oldest Fossils. ISBN: 9780764326974. $29.95

Library of Congress Control Number. 2011939221

Designed by Mark David Bowyer
Type set in Humanist 521 BT

ISBN. 978-0-7643-3880-9
Printed in China

Schiffer Books are available at special discounts for bulk purchases for sales promotions or premiums. Special editions, including personalized covers, corporate imprints, and excerpts can be created in large quantities for special needs. For more information contact the publisher:

Published by Schiffer Publishing Ltd.
4880 Lower Valley Road
Atglen, PA 19310
Phone. (610) 593-1777; Fax. (610) 593-2002
E-mail. Info@schifferbooks.com

For the largest selection of fine reference books on this and related subjects, please visit our website at **www.schifferbooks.com**
We are always looking for people to write books on new and related subjects. If you have an idea for a book, please contact us at proposals@schifferbooks.com

This book may be purchased from the publisher.
Include $5.00 for shipping.
Please try your bookstore first.
You may write for a free catalog.

In Europe, Schiffer books are distributed by
Bushwood Books
6 Marksbury Ave.
Kew Gardens
Surrey TW9 4JF England
Phone. 44 (0) 20 8392 8585; Fax. 44 (0) 20 8392 9876
E-mail. info@bushwoodbooks.co.uk
Website. www.bushwoodbooks.co.uk

Contents

Introduction

The early earth was not a place recognizable by standards of today's planet. Few land areas existed and those that did were barren of vegetation, no trees, grass or even more primitive land plants like mosses, liverworts or even lichens, the latter being a group of organisms which includes the photosynthetic cyanobacteria, a very ancient group of organisms. What did exist was bare rock with little or no regolith—bare rock which often bore the evidence of extraterrestrial impact. Land areas that existed were small—there were no continents. Continents have "grown" over geologic time from small, volcanic island-like landmasses known as island arcs to cover today a large portion of the globe. The earth's early atmosphere was anoxic (or reducing), no elemental oxygen was present. Carbon dioxide and water vapor made up most of the atmospheric gases. Volcanic activity was ubiquitous. Chains of volcano's associated with island arcs were fed by magma coming from the hot mantle. These volcanos poured out numerous flows of black, basaltic lava. Black lava would become (with deep burial and metamorphism) the greenstone of the greenstone belts of the Archean, some of the earth's oldest still existing rock. Unseen if an observer had been around at this early time would have been the granite and granite-like rocks, which were forming underneath parts of the greenstone belts. This stealth-like rock forming activity would continue on for a few billion years, being the main mechanism responsible for continent formation (and for keeping them above sea level).

The Archean oceans generally are believed to have been deep, with limited shallow water areas, so-called "littoral seas," which harbor the rich life of today's oceans. There were no (or few) continental shelves, which harbor much of the marine life that made its appearance during relatively recent geologic time (compared to the age of the earth itself), only a little more than half a billion years ago in the Cambrian period of the Paleozoic Era.

Chapter One

The Early Earth

The Bottom of the Stack

Geologic time spans some four billion years (the age of the oldest known rocks) to 10,000 years ago (the beginning of the Holocene and also of historic time). It's the earlier part of the earth's recorded rock record, which is presented in this book, a time and rock record that geologists refer to as the Precambrian. Precambrian time ranges from the age of the earliest Earth rocks (4+ billion years) to the Cambrian Period. The Cambrian Period being the oldest rocks containing **obvious** and (at least locally) **abundant invertebrate fossils**. The Cambrian is also the earliest period of the Paleozoic Era, with its beginning some 535 million years ago. Older rocks thus are **Pre-Cambrian** in age, in time as well as type, and represent a vast time span when the early evolution of life took place. Also in the Precambrian, the early evolution of the atmosphere took place. In the Precambrian's earliest part, the Archean, the last of the formative stages of the planet occurred.

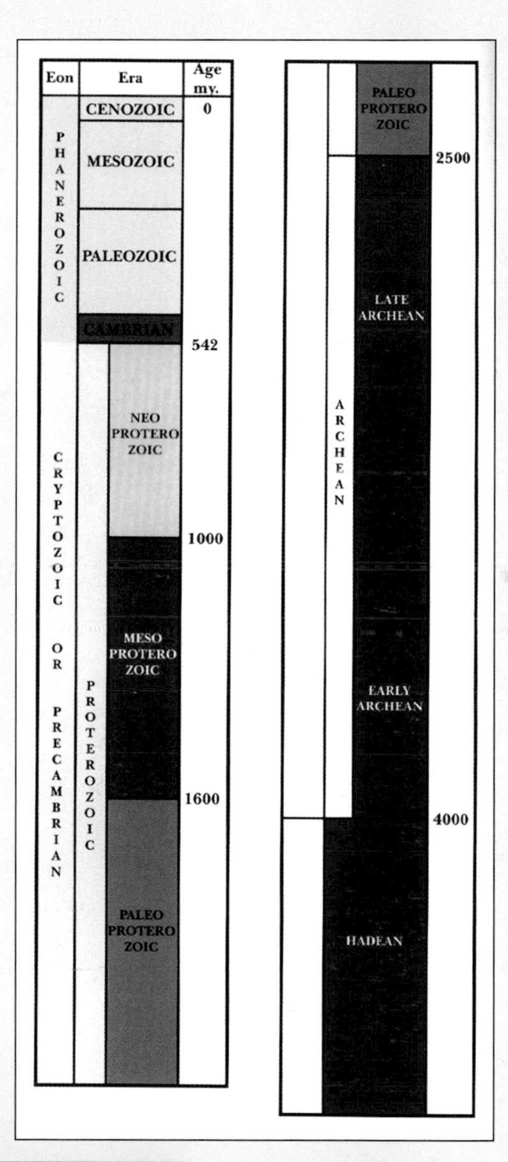

Geologic time scale. Most of geologic time consists of the Precambrian. The Precambrian is divided into two main parts, the Archean and the Proterozoic. Time before the Archean is known as the Hadean. The Hadean is **not** represented by rock records on Earth, but it is represented by meteorites and the rocks on some planets of the Solar System (Mars and Luna).

Geologic phenomena on the earth goes back over four billion years. This is almost one third of the time since the Big Bang, which was 13.7 billion years ago. In the geologic time scale, the Precambrian makes up over 80% of geologic time.

A rock, mineral, and fossil show. Throughout the states, rock shows have often replaced rock shops; brick and mortar institutions that have become less common than in the past. Shows such as this can offer a variety of geo-collectibles such as precious stones (cut and uncut), fossils, and meteorites, as well as minerals. Of the latter, geological age rarely is considered, a strategy that forms the basis of this work. Minerals, like fossils, can (sometimes) form a continuity through mega time.

A Planet of Basalt (The Black Earth Stage)

The earth prior to three billion years ago had no continents. Its crust was made of basalt, a black volcanic rock formed of material that cooled from a **magma ocean**, which had **existed half a billion years earlier**. This basalt surface saw extensive bombardment of meteors, the final stage in the accretion of the planet from primitive cosmic matter—matter that still can be found in the form of chondritic meteorites. This ancient black rock exists in parts of the planet's oldest crust, where it has been preserved from erosion as a consequence of deep burial, a process which sometimes changed its mineral composition to a rock known as greenstone. Appropriately, slivers of these ancient rocks are preserved today and known as **greenstone belts**.

Material of the mantle, below the basalt crust, remained in a hot and partially molten condition due to the presence of radioactive material present in amounts much greater than exist today, radioactive material that produced considerable amounts of heat upon decay. Basalt, when it came to the surface and cooled, became heavier than hotter molten basalt below, so that it would sink back into the mantle to be re-melted, thus being part of a convection cell. In this process, some elements, especially lighter ones like **silicon, sodium**, **potassium,** and **aluminum**, separated from denser ones and remained at the surface. This lighter, **felsic material** was not recycled back into the mantle. As a consequence, what developed was a series of small convection cells, with the molten basalt coming up and producing chains of volcanos. When it cooled and descended, it did so without the lighter felsic components, which were now being separated and **accumulated** over time at the surface. By this process, Earth's continents, with their concentration of lighter felsic material, were formed and enlarged.

Basalt, geologically modern, is essentially the same as basalt that has been coming out of the earth's mantle since the planet formed 4.5 billion years ago. Typical is the ropy surface formed when basalt comes out (extrudes) at the earth's surface or under water on the ocean floor. This basalt extruded on the surface of the land presents a surface that could have been seen anytime during the past 4.5 billion years, **except** for the land plants, which prior to 400 million years ago did not exist.

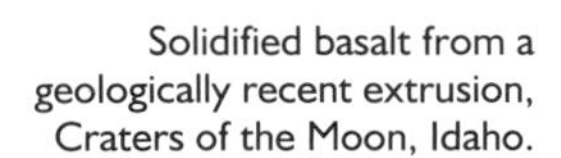

Solidified basalt from a geologically recent extrusion, Craters of the Moon, Idaho.

Ancient basalt (3.2 billion years old) from a greenstone belt in central Manitoba. Basalts are one of the major components in the oldest rocks of Earth, constituents of ancient regions known as "greenstone belts."

Archean pillow lava (basalt formed under water, left side) that has been intruded by granite (right), central Manitoba.

Archean basalt (bottom) overlain by pink granite that was intruded into the basalt and emplaced by underplating. Note the sharp contact between these two igneous rocks.

Greenstone. Originally black basalt but as a consequence of deep burial and metamorphism, it has had its black minerals changed to green ones. Basalts in greenstone belts can be converted to this green rock by metamorphism. Deep burial within the earth's crust, necessary to preserve rocks over long time periods from weathering and erosion, converts the basalt to greenstone—which is **green** like this!

Greenstone. Archean greenstone near Ely, Minnesota (Ely Greenstone). The green color of greenstone comes from iron bearing silicates. When these weather, they produce rusty looking iron minerals that discolor the greenstone, as has taken place here. If a fresh rock sample is broken from the outcrop, it would be green.

Close-up of greenstone from the outcrop shown in previous photo.

Green greenstone! Weathering oxidizes the iron in greenstone to ferric oxide, which can then infiltrate along cracks in this hard rock.

Precambrian and Phanerozoic Rock. Here Precambrian granite is overlain by younger sedimentary rock, which still is some 500 million years old. This younger rock, which can contain obvious fossils, belongs to what is known as the Phanerozoic, that time span postdating the Precambrian. Precambrian rocks, like this granite, have long been recognized as being very ancient. At one time such ancient rocks were considered to have formed when the planet itself cooled. It is now known that granite is generally geologically ancient but is not the earth's original crust. It formed as a light, lower melting point fraction derived from primitive mantle material—material added to the crust from below (under-plated), building the continents. To expose such granite at the earth's surface takes hundreds of millions of years of erosion and weathering. Those rocks that were above the granite had to be to removed to expose the granite, rocks that might have been greenstones. The semi-circular masses in the granite formed as a consequence of weathering, possibly from ancient weathering that took place before the much younger overlying layered sedimentary rocks were deposited.

Mafic (black) dikes in granite. Black rock (forming the bands) was injected into cracks in the lighter granite about one billion years ago. This would mean that the granite is older than the mafic rock. Mafic rock comes from more primitive magma than does the lighter colored granite—the magma which formed the granite having been derived from lower-melting-point and lighter components which were also derived from the mantle.

The same dike as shown in the previous photo at a nearby location. Here the mafic rock has been deeply weathered and has been converted to clay. The rock roof above the man is composed of cemented-together-boulders, which laid upon the surface of Missouri some 700 million years ago. Above this conglomerate layer and thus slightly younger than it are sandstone beds containing fossils of some of the earth's oldest shelled animals, brachiopods, which lived in shallow seas some 530 million years ago.

Vertical mafic dike injected (intruded) into 1.5 billion year old felsic igneous rock of southern Missouri. The holes to the left of the girl are from the taking of oriented samples used in the determination of paleomagnetic vectors used to determine continental drift.

Typical river scene where a stream has cut into hard Precambrian rocks. This is on the **Canadian Shield** in northern Minnesota.

A typical Precambrian outcrop—often rocky and scenic. Granite on the St. Francois River, Silver Mines, Madison County, Missouri Ozarks.

Another view of typical rocky Precambrian terrain in the same area as the previous photo (Proterozoic granite).

The Bottom of the Grand Canyon

Deep in the inner gorge of the Grand Canyon occur ancient rocks which, to most persons, give reality to the great age of the earth. What most don't realize is that ancient rocks like these can occur at the surface of major portions of the planet, especially in those areas **known as shields**.

This view looking down into the inner gorge of the Grand Canyon shows Precambrian rock cut by the Colorado River. The dark colored (and non-layered) rock of the inner canyon is the mid-Precambrian Vishnu Schist. Overlying this metamorphic rock is a mile of sedimentary rock (which forms the upper half of the picture) belonging to the Paleozoic Era. The fact that the rock at the bottom of the canyon is extremely ancient is somewhat obvious, considering the time required to deposit the overlying rocks and then for the Colorado River to cut the canyon. There is also what is known as a profound unconformity (a hiatus) between the Precambrian metamorphic rock and the overlying sedimentary layers; just above this is the so-called Tonto Platform, a flat terrain developed on horizontal Cambrian sandstone.

View of the inner gorge of the Grand Canyon, Arizona. These are metamorphic rocks upon which sedimentary rocks of various ages have been deposited to make up the rest of the canyon. Notice the fault in these sediments at the left side of the photo.

Looking into the lower half of the canyon made up of the Precambrian Vishnu Schist. The Vishnu Schist forms much of the inner gorge of the canyon.

Lower half of the Grand Canyon exposing Precambrian Vishnu Schist. Horizontal beds just above the non-layered schist belong to the Cambrian Period of the Paleozoic Era.

The Vishnu Schist exposed in the steep walled inner gorge of the Grand Canyon. Precambrian rocks, especially crystalline examples like the Vishnu Schist, are much harder than are the younger rocks of the upper canyon. The Colorado River has had a more difficult time cutting its gorge into these hard, ancient rocks. It has not been able to cut laterally into the schist.

Bridge on the Bright Angel Trail crossing the Colorado River. This is at the bottom of the Grand Canyon—over a mile below the Kiabab Plateau. The Vishnu Schist is a high grade metamorphic rock of early (or mid-) Precambrian age extensively exposed in the inner gorge of the Grand Canyon.

Gneiss is the rock forming some of the cliffs of the inner gorge of the Grand Canyon. It's a common rock often characteristic of the Archean.

The same bridge shown above is in the background with the Bridge over Bright Angel Creek in the foreground. All of this is in the inner gorge of the canyon—that portion of the canyon cut into ancient basement rocks.

These peculiar, sausage-like structures (boutanage) are in another high grade metamorphic rock (gneiss) of the inner gorge of the Grand Canyon. The girl is pointing to these structures, which are characteristic of rocks that have been deeply buried in the earth's crust and then subjected to high pressures (the pressures of mountain making). These mountain ranges existed during the Precambrian, **but were completely eroded away before the overlying sedimentary rocks of the canyon were deposited.**

Highly fractured gneiss—inner gorge of the Grand Canyon.

Hot spring deposits, Yellowstone National Park. Early rock records give considerable (indirect) evidence for the existence of a great deal of hot spring and geothermal activity. The young earth was thermally and tectonically more active than it was during later geologic time (or is today). This is because greater amounts of radioactive material existed in the early earth. This elevated radioactivity produced greater amounts of heat energy, the energy source of hot springs, and also the force that drives tectonic activity. The brown, yellow, and red colors of the slimy hot spring surface comes from various primitive life forms, which herald back to the early earth. These can include the thermophilic archaebacteria, a domain of primitive organisms existing under extreme conditions of both high heat and salinity or chemically toxic conditions.

Sapphire Geyser, Yellowstone Park, Wyoming. Geysers were common and widespread features of the early earth. Today they are rare except in limited areas like Yellowstone.

Erupting geyser (at right in background) and hot spring deposits, Yellowstone Park, Wyoming. Geothermal activity of the early earth looked like this, but of course without the trees or other vegetation seen in the background.

Geyser pool. Geysers and hot springs are fed by underground conduits that heat and conduct groundwater. Here a geyser is just beginning to erupt. *Photo courtesy of Warren Wagner.*

Initial eruption of Sapphire Geyser, Yellowstone Park, Wyoming.

Geyser pool that leads to a geyser conduit.

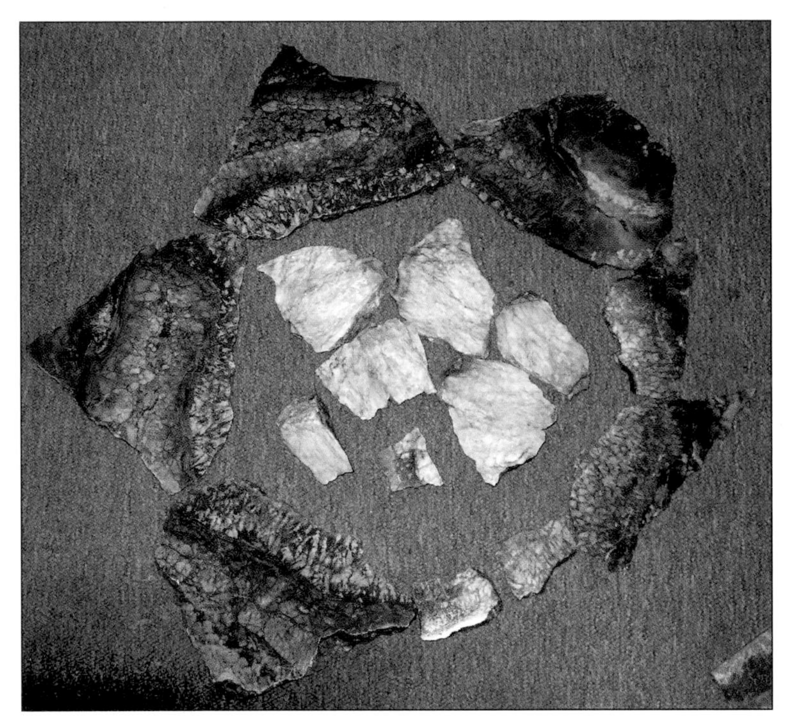

Reconstructed hot spring or geyser conduit from 1.5 billion years ago. Excavations for Highway 67 near Fredericktown, Missouri, encountered a number of more-or- less circular tubes filled with this reddish calcite. These appear to be the mineral-filled conduits of either hot springs or geysers. Geysers in Missouri? Yes, a billion and a half years ago there was much volcanic activity in the US Midwest. Where there is a lot of volcanic activity, there can be geysers. Unfortunately, these mineral filled tubes were removed or totally covered when the road cut was completed.

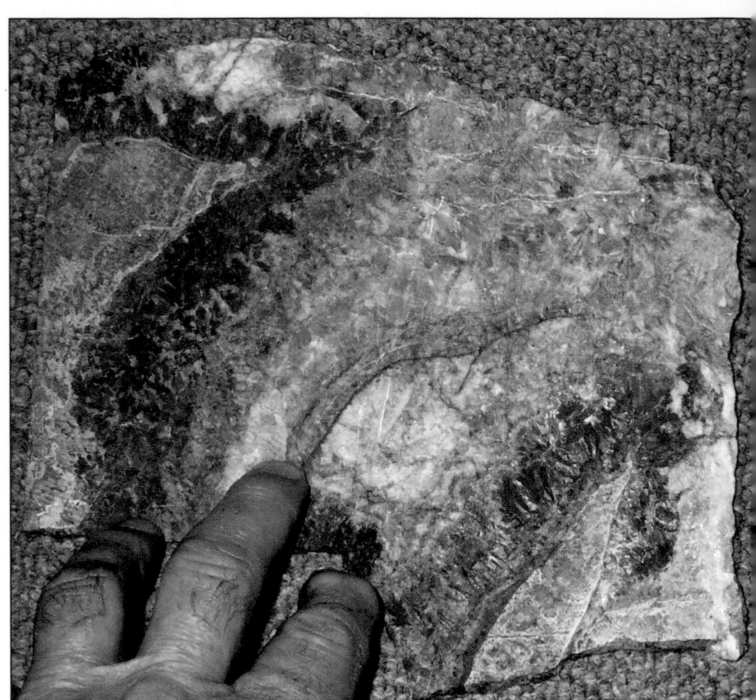

Close-up of hot spring slab.

Road cut on Highway 67 cut into volcanic ash (rhyolite), which contained the above mineral-filled tubes.

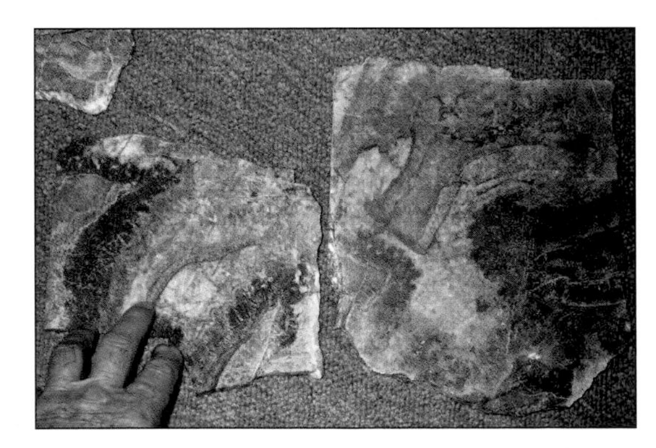

Slabs of (presumed) hot spring or geyser deposits from the previously mentioned locality. (Value range E).

Geothermal Phenomena

Hot springs, geysers, and other geothermal phenomena associated with volcanism were common and widespread in the early earth. Radioactive material, the major energy source of geothermal activity, existed in amounts greater than exist today and thus could produce extensive heat-dependent-activity like hot springs and geysers. Phenomena shown here could commonly have been seen in the Archean, except for the trees, grass, and blue sky. (The blue comes from elemental oxygen, which was absent in the atmosphere of the early earth.)

Impact Phenomena

Besides geothermal activity, meteoroid impacts were the other ubiquitous phenomena of the early earth.

Archean impact breccia? Rocks composed of angular fragments like this were derived from different sources and are seen in the early earth rock record. Some of these have been explained as being impact breccias formed from extensive meteor impacts, which took place on the early earth. This breccia is some three billion + years old. It came from a greenstone belt that formed at the same time as did many of the craters on the Moon and Mars. On Earth, few geologically ancient craters survived. These ancient cratered surfaces have either been removed by erosion or destroyed by other means. Earth is a geologically active planet and three billion years is a long time, even for geology.

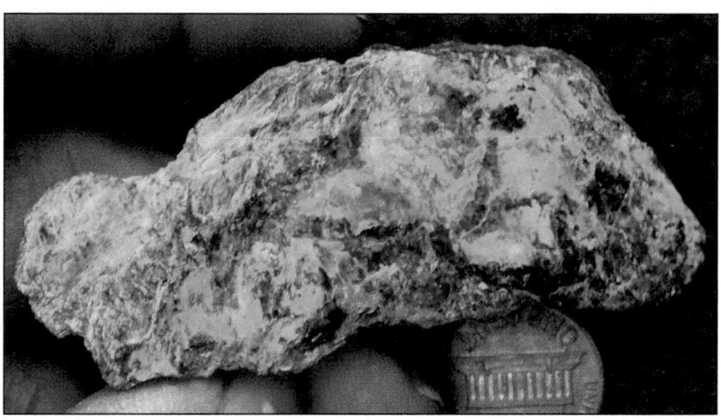

Zaratite. A rare nickel mineral from early Precambrian rocks of Tasmania (Lord Brassey Mine). Nickel minerals are sometimes associated with rocks of the early earth. Numerous asteroids and meteorites that bombarded the early earth contained nickel. Meteorites sometimes were assimilated into the early crust, forming the nickel minerals.

Chrysoprase. This rock, often utilized in lapidary work, gets its greenish color from nickel compounds. It is associated with very ancient rocks of the Archean. This chrysoprase is from Australia, a continent that includes extensive outcrops of rock formed in the Archean Era, rock representative of the early earth. From Yerilla, Western Australia, 100 miles north of Kalgoorlie. (Value range H, single specimen).

Early Minerals

This book is primarily about collectable minerals—**but collectable minerals with a focus**. The author is a geologist who has a bent for and focus on fossils. This paleontological focus on the early earth has been manifest in his Schiffer book, *The World's Oldest Fossils*. Occurrences of fossils have long been recognized to occur in rocks of the earth's crust, where they **follow a sequence related to time**, or in geologic terms they are in a stratigraphic sequence. (It is for this reason that fossils are used to place strata in a time sequence and also why they form the basis for the geological time scale). **Minerals**, on the other hand, **rarely are referenced in this way**, in a **context of geologic time**. Rarely does a mineral discussion include the ages of the rocks from which the minerals were collected or occur. The approach here is different! Minerals that formed early in the earth's history are focused upon. Those minerals and mineral occurrences associated with the older rocks of the earth's crust are especially favored. In some cases, this is because these minerals and rocks reflect conditions that existed at that early stage in the geologic history of the earth, conditions unique to that time and might seem to violate geology's "principle of uniformitarianism."

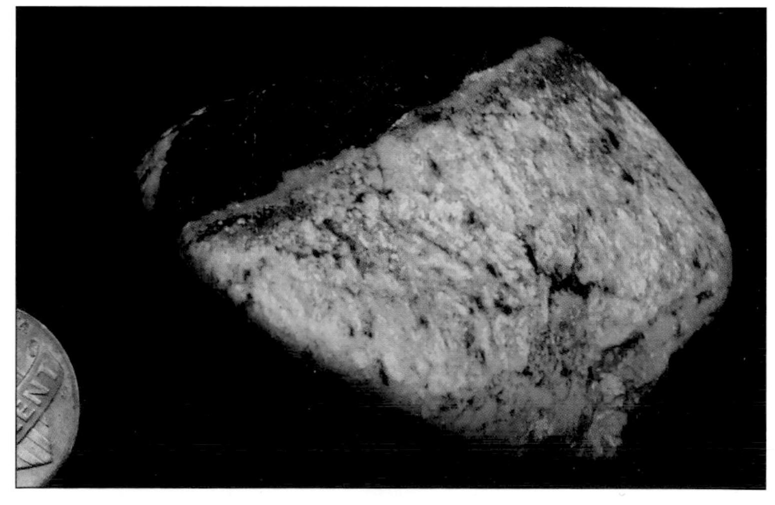

Close-up of chrysoprase polished pebble with a granular texture.

What has to be considered, however, is that the earth, like life itself, has undergone a profound evolutionary process over mega time.

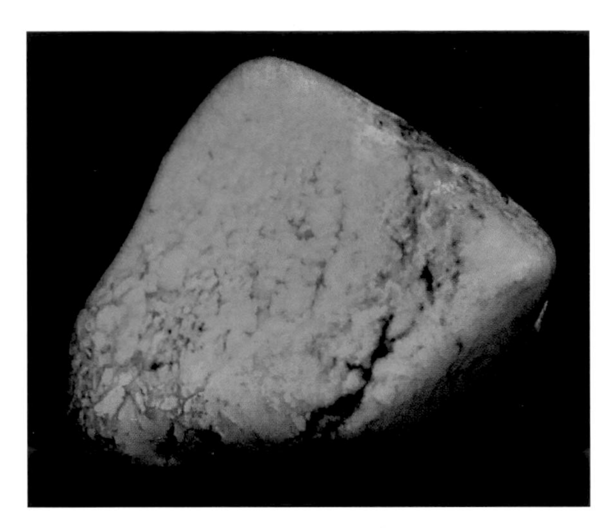

Chrysoprase pebble. This is one type of "Ancient Earth Phenomena," which has become widely distributed among rockhounds.

THE EOZOIC AGES.

THE dominion of heat has passed away; the excess of water has been precipitated from the atmosphere, and now covers the earth as a universal ocean. The crust has folded itself into long ridges, the bed of the waters has subsided into its place, and the sea for the first time begins to rave against the shores of the newly elevated land, while the rain, washing the bare surfaces of rocky ridges, carries its contribution of the slowly wasting rocks back into the waters whence they were raised, forming, with the material worn from the crust by the surf, the first oceanic sediments. Do we know any of these earliest aqueous beds, or are they all hidden from view beneath newer deposits, or have they been themselves worn away and destroyed by denuding agencies? Whether we know the earliest formed sediments is, and may always remain, uncertain; but we do know certain very ancient rocks which may be at least their immediate successors.

Deepest and oldest of all the rocks we are acquainted with in the crust of the earth, are certain beds much altered and metamorphosed, baked by the joint action of heat and heated moisture—rocks once called Azoic, as containing no traces of life.

2*

Perceptive interpretation of events of the early earth by James W. Dawson from the "Story of the Earth and Man", 1873. J. W. Dawson was one of the pioneering geologists of Canada. The term Eozoic was a term used for the early Precambrian and its disturbed rocks. It is the same in most ways as is the Archean of today.

Gniess and Other Early Rock Phenomena

Many of the ancient specimens illustrated here have shown up at mineral and rock shows over the past two decades as geo-collectables related to the **early earth**. Robotic exploration of planets and satellites of our Solar System during these same two decades have given a much clearer picture concerning the origin and evolution of the Solar System. The uniqueness of the earth, not only in terms of its biosphere but also in how the biosphere itself has interacted with and influenced the earth's evolution, is a major consideration. As a consequence of space exploration, evolution through three billion years of geologic time has become better appreciated and clearer.

Gneiss. Banded gneiss is one of the most common rock types representative of the early earth. It formed from deep burial and metamorphism of either granite or sedimentary rock like greywacke, siltstone or shale—all rocks high in clay minerals. This outcrop is in western Ontario, on the Canadian Shield.

Close-up of previously shown outcrop. Gneiss is sometimes said to be formed in the "roots of mountains"—**mountain ranges that no longer exist**!

It is now realized that much of the geological evolution of the earth has been guided and influenced by phenomena associated with the biosphere, phenomena like photosynthesis, with its production of free, elemental oxygen being among other important considerations. This book is focused toward minerals and related geochemical phenomena, but with a bent toward those geochemical processes dependent upon the **phenomena of life**. Thus, chapter five deals with biogenic sediments and sedimentary rocks influenced by the biosphere, specifically iron formation. More obvious with regard to the earth's biosphere, chapter six deals with fossils, specifically Precambrian fossils like stromatolites and the puzzling Ediacaran biota of the late Precambrian. Chapter eight concerns itself with carbonate rocks (and their associated minerals), especially the minerals formed from carbonates when they were extensively "cooked" to form beds of crystalline marble.

Gneiss with numerous quartz veins and aplite dikes. Superior Province, western Ontario, Canada. Archean rocks, usually having been deeply buried within the crust, are often full of quartz veins like this, which indicate a lot of tectonic activity. These rocks were at one time at the "roots" of mountains.

Gneiss with granite dikes (aplite dikes) and quartz veins. Kenora, Ontario.

Volcano's and the Black Earth Stage

Volcanos (a spectacular form of geothermal activity) formed from the rising of mafic magma that created narrow **volcanic islands known as island arcs, which were predominant during the early earth.** This time period of extensive mafic volcanic activity constitutes the "Black Earth Stage" of the earth. Black basalt in the crust formed from both the cooling of the earth's Hadean "magma ocean" and from basalt of the early island arcs. Felsic material that separated from basalt became the nuclei of continents, which enlarged over time with the addition of more felsic material, which eventually would become **continental crust**. As the convection cells operated, a fraction of this lighter felsic rock—that rock enriched in potassium, sodium, aluminum, and silicon (**felsic material**)—ascended and either "plated" the mafic rock from below or was injected into it.

Large numbers of quartz veins in grey gneiss indicate a great deal of hydrothermal activity when these rocks were once part of an extensive mountain range. This situation is indicated by the quartz veins, a common feature of rock formed during the early earth. Trans-Canadian Highway, Kenora, western Ontario.

Homogeneous Archean gneiss, Rocky Mountain Park, Estes, Colorado.

As more of this lighter material separated from primitive mantle material, it became depleted of lighter elements which, when they came to or near the surface, encased more and more of the island arc basalt. Lighter material encased at least those portions that did not get carried back into the mantle (by the process known as subduction), thus preserving the lighter matter. This lighter-in-density and lighter-in-color felsic material continued to come up, with massive amounts being "plated" under parts of the crust known today as batholiths. These batholiths (which usually are composed of large masses of granite) can now be at the earth's surface where they can form mountain ranges because of granite's resistance to weathering and erosion.

Greywacke and Mafic Volcanic Rock

At the same time as this granite **under plating** was taking place, basaltic volcanic rock was being subjected to a carbon dioxide atmosphere, undergoing weathering. From the weathering of this dark rock was produced dark, iron-rich sediments consisting of "dirty sand" mixed with fragments of other mafic volcanic material. Sometimes this dirty sediment became greywacke (a type of sedimentary) rock. When greywacke gets deeply buried in the crust (necessary to preserve it over vast periods of time), it gets very hard and may be difficult to distinguish from its parent basalt. Thus, many of the greenstone beds of the earth's earliest crust will be found composed of hard black (or green) rock. It cannot be easily determined whether these rocks originally were basalt or whether they were the ("cooked") sedimentary rock greywacke.

Metamorphosed greywacke (dirty sandstone) with quartz veins, Nonacho Lake, NWT Canada. Sand (and rock made from what was once sand, like this rock) of the early earth was dirty. It had in it, besides quartz, a lot of material such as basalt fragments and clay. Over vast spans of time, such "dirty" sand has been recycled many times over. Each recycling "cleaning it" a bit so that the sand of later geologic time (and today) is "clean" and composed predominantly of quartz. The other material in the sand having been removed through its many cycles of cleaning.

Quartz veins in schist or meta-greywacke, Mt. Zirkel, Colorado. Archean sediments, especially sand-like sediments, were "dirty"—"clean" quartz sand like that commonly seen today was rare. Clean sand is a product of long periods "recycling" over and over. Geologic recycling is a cleaning process. The "dirty" sand of the Archean is often the parent material of metamorphic rock like this, metamorphic rock that is often shot full of quartz veins.

Uniqueness and Occurrences of Early Earth Minerals and Rocks

Minerals and rocks of the early earth sometimes are different from those of later geologic time. This appears to be in violation of geology's principle of uniformatarianism. Different tectonics and the lack of (or minimal) sorting of trace elements resulted in conditions that were different from those of more recent geologic time. Records of these early events are found in the center of uplifts in what in Europe are known as massifs. Massifs are places where rocks, normally found deep beneath younger rocks, have been **brought to the surface as a consequence of uplift**—uplift accompanied by the erosion of the younger, overlying rocks over hundreds of millions of years. Some areas of the US where this has happened include large parts of the central Rockies of New Mexico, Colorado, and Wyoming. Others are the Llano Uplift of Texas, the Ozark Uplift of Missouri, and the Black Hills of South Dakota. Ancient uplifted rocks are also found forming the cores or central parts of the Appalachians as well as parts of western Montana, Utah, and Arizona, including the **inner gorge of the Grand Canyon**. The most extensive areas of these ancient rocks occur in what are known as **shields,** which are generally broad, relatively low regions. The largest shield on the planet is the Canadian Shield, which covers one third of Canada and major portions of the states of Minnesota, Wisconsin, and Michigan. The Canadian Shield also covers most of the upper portion of Quebec (New Quebec), much of western Ontario, the northern parts of Manitoba and Saskatchewan, and much of Northwest Territories, including the (recently established) territory of Nunavut in arctic Canada. Also in the northern hemisphere, the Fenno-Scandinavian or Baltic Shield forms all of Finland, eastern Norway and portions of Sweden and northwestern Russia (Kola Peninsula). Other sizeable shield areas occur in Asiatic Russia (Angaran Shield), and in large portions of Africa (which includes the Ethiopian Shield). In South America, the Amazonian or Brazilian Shield forms the eastern bulge of South America. This is the source of many of the beautiful minerals covered in chapter three. North of this is the Guinean (Venezuelan) Shield and to the south, the Plabian Shield. Large shield areas also exist in the western half of Australia, much of the southern Indian Peninsula (Indian Shield) and in China, especially that area bordering China and North Korea.

Granite spire weathered from part of a batholith. Granite usually forms from felsic magma that "under plated" or was added to the crust from below by rising felsic material—material previously separated from mantle material. Exposed much later on the surface through weathering and the removal of considerable thicknesses of once overlying rock, granite can then be carved by erosion into a variety of shapes as seen here. City of Rocks, near Albion, Idaho.

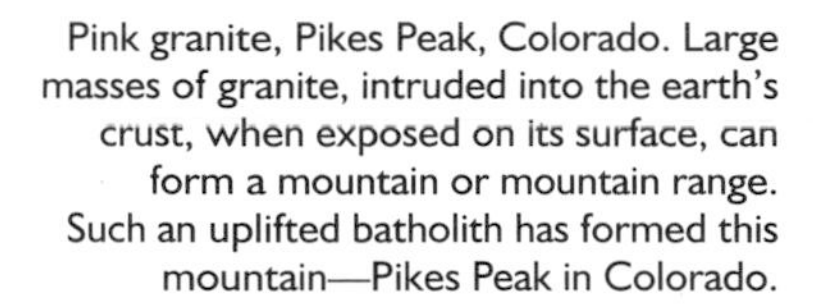

Pink granite, Pikes Peak, Colorado. Large masses of granite, intruded into the earth's crust, when exposed on its surface, can form a mountain or mountain range. Such an uplifted batholith has formed this mountain—Pikes Peak in Colorado.

Pegmatite. Granite batholiths can contain pockets of large crystals of the same minerals that compose the granite. Such pockets of large crystals are known as pegmatites and can be the source of gem mineral crystals like topaz, beryl, and aquamarine.

Entrance to a mine in a weathered pegmatite in Minas Gerais, Brazil. This is a typical entrance to a small mine working a gem pegmatite. The pegmatite is located in deeply weathered granite where chemical weathering taking place in a tropical climate has released and enhanced the silicate gem minerals.

Ancient Granitic Rocks in Tropical Regions

At low latitudes, especially in the Brazilian and Guinean (Venezuelan) Shields, deep weathering under tropical conditions has converted what otherwise would be very hard rock like granite and gneiss into thick beds of clay—rich, red regolith. **The granite batholiths of these areas contain masses of large crystals known as pegmatites.**

This deep weathering allows resistant-to-weathering minerals' release from these hard rocks. These weather resistant minerals are often prized as gemstones. Thus the tourmalines, aquamarines, chrysoberyls, and other hard silicate minerals occurring in pegmatites can be dug from what is now clay, rather than extracted from hard rock, as would be the case in the higher latitude shield areas. Similarly to South America, shield areas existing in sub-Saharan Africa produce what is known as laterization, the formation of thick, red soil. This soil occurs where deep chemical weathering has taken place in a tropical environment.

The large shield regions of Australia, by contrast, usually find these ancient rocks relatively fresh and unweathered as a consequence of the dry climate. Other regions of very ancient shield-like rocks are found in Madagascar and Greenland, although these regions (islands?) can be considered as extensions of the shields of Africa and North America respectively.

Tourmalated quartz crystal, Minas Gerais, Brazil. This is a quartz crystal containing tourmaline. It came from a pegmatite emplaced in granite that was deeply weathered in a tropical environment. (Value range F).

Ancient rocks containing pegmatites are also well represented in China, and smaller shield fragments contain them in parts of Afghanistan and Pakistan. The latter are in highly uplifted and geologically disturbed portions of the Asian landmass. These regions also produce gem minerals from pegmatites, as do parts of India, Burma, and Sir Lanka, areas which have been famous for gem stones for millennia.

Tourmaline crystals embedded in quartz. This mineral (or gem) specimen came from pegmatites that were deeply weathered in the tropical environment of Brazil, the parent granite forming a part of the large Brazilian Shield. Pegmatites that produce minerals like this and that weathered under tropical conditions are known as **gem pegmatites**.(Value range F).

Rutile crystal, Brazil. Rutile (titanium dioxide) is also one of the minerals associated with gem pegmatites. It occurs in pegmatites of other regions also, but because it is so tightly held to associated minerals, it usually is not able to be profitably extracted.

Faceted tourmalated quartz, Brazil.

Kunzite. A variety of the pegmatite mineral spodumene, kunzite is one of various lithium minerals associated with gem pegmatites. (Value range F).

Ancient Rocks of the Far North

All of the regions mentioned here have very ancient rocks, but the large Canadian Shield has rocks that are some of the oldest of the old. What is known in Canada as "The Shield" is composed of a number of geologic sub-regions, some of which, like the Grenville Province, have been recognized as a consequence of their distinctive geology since the mid-nineteenth century. Other subregions of the Canadian Shield have been recognized more recently, not only from their geology, but from a similarity in radiometric age dates, some of which have been found to be very ancient. One of the oldest portions of the shield (determined by radiometric age dating) is that portion at its southeastern edge that also includes Greenland. This is known as the Nane Province and includes both Labrador and southern Greenland. Rocks of the Nane Province have been found to be some of the oldest on the shield. And some of the rock types occurring in the Nane Province also occur in other ancient portions of the earth—regions whose rocks have also determined as being very ancient.

Labradorite

One of the rock types associated with this very early crust is composed of the mineral labradorite, a calcium-sodium feldspar showing a beautiful play of iridescent colors. The Nane Province also includes rocks of the Isua Series of Greenland—an area of radiometrically dated rocks that approach four billion years in age. The rocks of the Isua Series represent some of the oldest rocks on the planet. Possibly even older rock of the Canadian Shield is the Acasta Gneiss (if age dating by radiometric methods of their enclosed zircon crystals is accurate), the Acasta Gneiss **is** four billion years old. This rock occurs in the northwestern part of the Canadian Shield in NWT (Northwest Territories), Canada. Similarly aged gneiss occurs east of Hudson's Bay in New Quebec and individual zircons from Australia have given age dates of over 4.2 billion years, the only actual earthly geologic specimens from the Hadean. Other occurrences of labradorite, identical to that of Labrador, occur in Madagascar. The Madagascar occurrence, with its sequence of Greenstones and labradorite, may represent a fragment of early crust that separated from the Nane Province sometime during the Precambrian. The two identical sequences now being half a world apart.

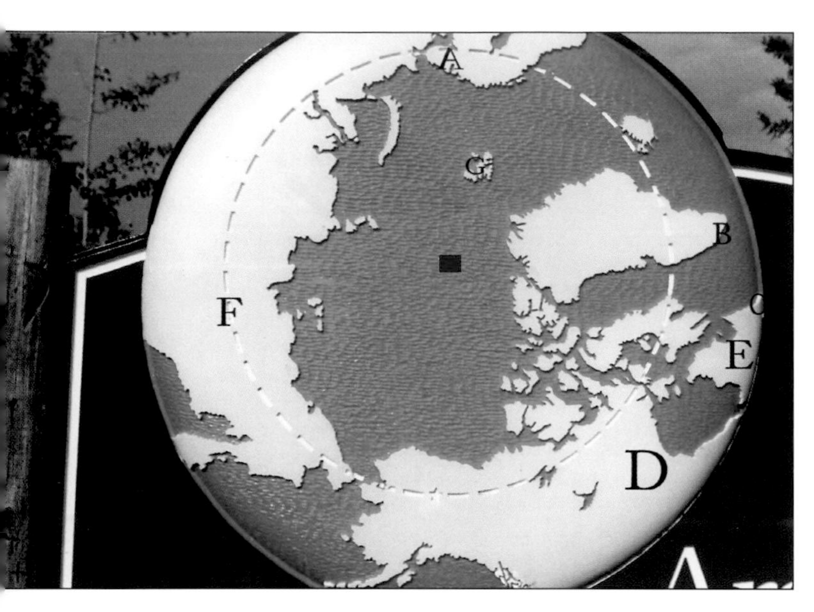

The earth viewed from the north pole. Part of an interpretive sign on the Dalton Highway (Alaskan pipeline highway) where it crosses the Arctic Circle. The letters represent the location of the following outcrops of early Precambrian (Archean) rocks on the Canadian Shield. A is the Kola Peninsula of Russia, B the Isua Series of the Nane Province of Greenland, C represents occurrences of Labradorite in the Nane Province of the Canadian Shield of Labrador. The letters indicate outcrop areas of these very ancient rocks discussed. D and E are in the northern part of the Canadian Shield, F is the Aldan Shield of Russia. Nature places many areas of the oldest parts of the planet in remote places, many in the far north.

Labradorite from Labrador. This is a feldspar occurring in rocks of the Nane Province of the Canadian Shield. Labradorite crops out along the coast of Labrador close to the area of the previous photo.

Waterfall in Labrador. This is typical scenery of the Canadian Shield. Rock forming this waterfall is granite of the early Precambrian Nane Province of the eastern portion of the Canadian Shield. Large portions of "The Shield" consist of wilderness with few roads and lots of lakes, mosquitoes, and black flies.

Labradorite cleavage surface with its typical play of colors. Specimen from Labrador. (Value range G).

Polished Labradorite from Labrador. (Value range F).

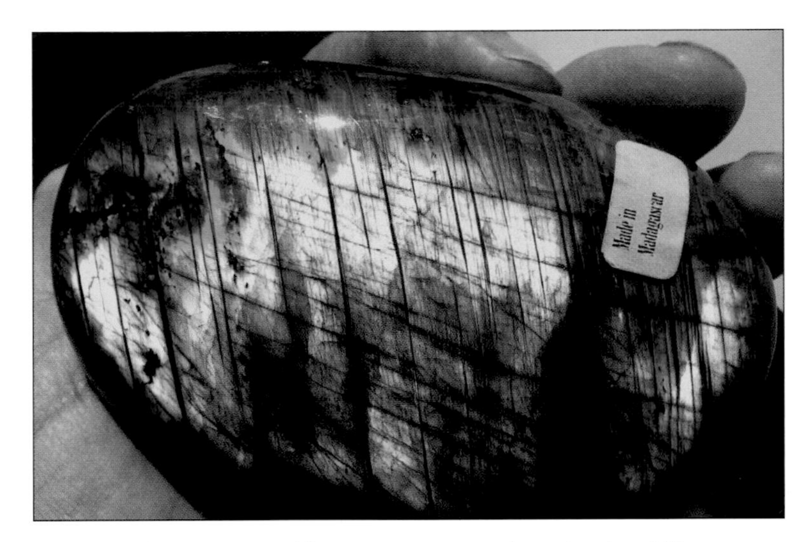

Labradorite. Same specimen, different position—note change in color of this feldspar. Labradorite's color comes from diffraction, the same phenomena that gives a compact disk its play of colors. Labradorite from Madagascar is identical to that of Labrador. Both are very ancient, from the early Precambrian or Archean Era. Plate tectonics has shown that portions of the earth's crust can be transported great distances and then incorporated into continental crust of a distant region. The labradorite found in Madagascar may have come from the same source (gabbroic batholith) as that found in Labrador. A sliver of Labrador's labradorite, transported around the globe by ancient sea floor spreading, was added to what is now Africa (Madagascar has long been recognized to be geologically a part of Africa).

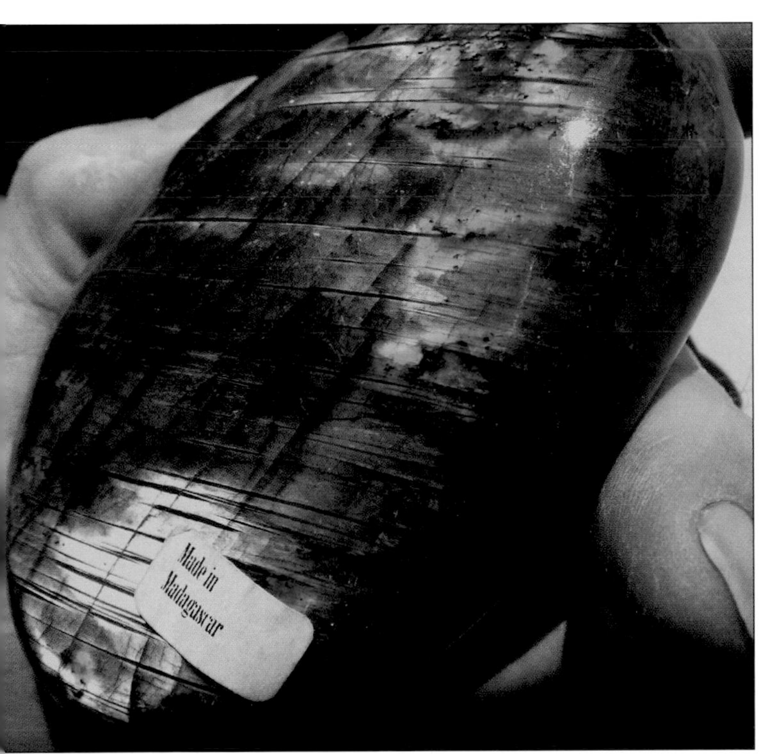

Labradorite seen here is identical to that from Labrador but is from Madagascar. Labradorite from Madagascar has been widely distributed among mineral collectors and rockhounds. It may well be from the same intrusion as that of Labrador as plate tectonics has demonstrated that pieces of the globe can be transported great distances and then incorporated into continental masses distant from where they originated.

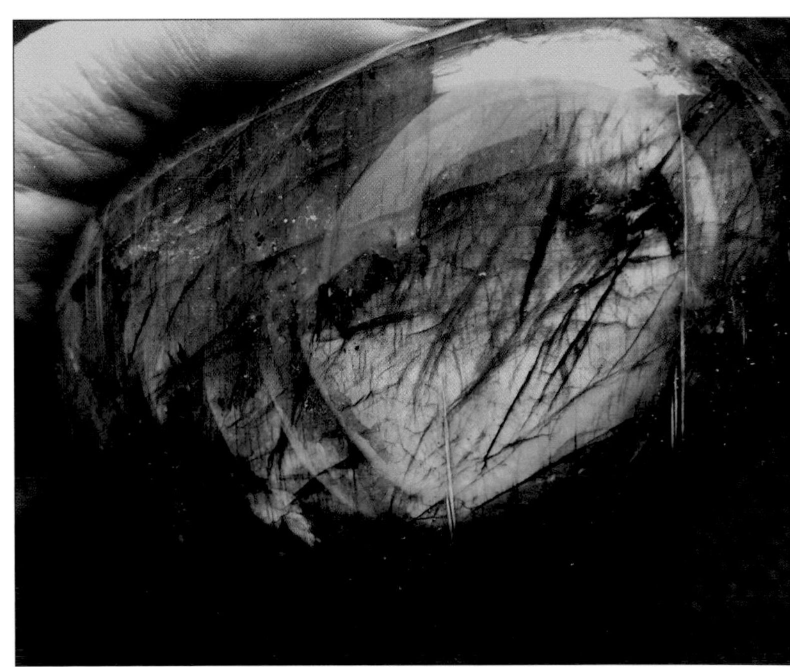

Labradorite from Madagascar. The labradorite from Madagascar, like that of Labrador, is very ancient.

Geologic Records of the Far North and the Early Earth

The far north has been, and still is, a frontier that has contributed extensively toward geologic understandings of the Archean and the early earth. The northern portion of the Canadian Shield, being a source of uranium minerals (Great Slave Lake Region, NWT and northern Saskatchewan), is not only a major source of nuclear energy (with its uranium) but also a source for understanding mega time through radiometric age dating of its ancient rocks (the discipline of geochronology). Other portions of "the shield," in northern Quebec, Labrador, and Northwest Territories, have yielded both mineral wealth and a wealth of geologic information on the Archean earth (including Nunavut, that portion of NWT that became a specific Canadian Territory to recognize the claims of the Inuits or Eskimos). All of these regions represent large areas of Precambrian terrain, some portions of which offer some of the planet's oldest (and still only partially understood) ancient geologic records.

Other portions of the far north that have recently contributed toward an understanding of the early earth are the islands of Svalbard (Spitsbergen), which occur in the far North Atlantic-Arctic ocean. Here sequences of Proterozoic sedimentary rocks have yielded biotas of microfossils and stromatolites. These artifacts reinforce our understandings of the fossil record of life about a billion years ago. Archean records on the Canadian Shield include the gold bearing greenstone belts of NWT near Yellowknife and Timmins, Ontario. These areas are both economically and geologically significant, which makes Canada one of the planet's largest producers of gold.

On the other side of the planet, the ancient (and geologically peculiar) terrain of the Kola Peninsula in northern Russia represents another part of the far north that has contributed to the understanding of the early earth. All of these areas have had some of their geologic secrets unlocked with the development of means of transportation capable of accessing and traversing this once almost inaccessible terrain. Aircraft, especially the float plane and the helicopter, have significantly aided in this exploration.

Chalcopyrite, Nonacho Lake, NWT Canada. Chalcopyrite is a copper-iron-sulfide. It's a valuable and important ore of copper. These minerals occur associated with quartz veins formed in Archean greywackes.

Exploration trenches near Uranium City, Saskatchewan, on the border of NWT and northern Saskatchewan (60 degrees latitude). Parts of the Canadian Shield like this, near the Saskatchewan-Northwest Territories (NWT) border contain extensive deposits of pitchblende (uranium oxide). Mineral exploration and production like this has made parts of these shield areas and their geology accessible.

Soapstone carving. Archean talc schists of the northern part of the Canadian Shield have for many years been carved by the Inuits to make interesting carvings. These carvings were encouraged by the Hudson's Bay Company as an income generating craft for the Inuits.

Field camp on "The Shield." A typical summer "Bush Camp" set up to do exploratory geologic mapping and mineral prospecting. Large portions of the earth's Precambrian shield areas are still not well known geologically. They are geologically complex and are often in remote areas like this.

The Far North's Geology and the Importance of Transportation

Exploration of the varied geology of the far north would not have been possible without aircraft, especially the float plane and the helicopter. These technologies offer the ability to explore the region of the far north with less travail than was previously possible. Mineral wealth and mineral resource development have also benefited and developed here in part thanks to the same types of aircraft. Not part of the early earth picture, but significant in far north development and access, are the petroleum resources of Prudhoe Bay in Alaska and Norman Wells in NWT Canada. This resource development contributed materially toward access of the far north and aircraft helped to open it up for acquisition of its mineral resources. Aircraft have allowed for relative ease of access to these far locations, especially when compared with 100 years ago, when the reaching of the North Pole was a goal achieved by the use of sled dogs on the surface of the frozen Arctic ocean. Today, commercial flights between North America and Asia fly over the polar region, a region accessed only for the first time in 1909.

Precambrian outcrop in Alaska. Slivers and masses of Precambrian rocks exist in Alaska. These appear to be slices of Precambrian age crustal rocks derived either from the edge of the Canadian Shield or from elsewhere and later incorporated into the complex of rocks composing much of Alaska—the so called autochronous terrain.

Float planes can take off and land on the many lakes of the north, especially in glaciated areas on "The Shield." This arrangement has enabled access to large portions of the far north.

Kola Peninsula mineral. One far north region of European Russia (that part of Russia west of the Urals) known as the Kola Peninsula, is a continuation of the Fenno-Scaninavian Shield of Finland and Sweden. The Kola Peninsula exposes some of the earth's oldest rocks, some which are mineralogically unique. The Kola Peninsula is also the site of the deepest hard rock drill hole, a deep crust sampling project initiated by the Soviet government in the 1980s to systematically sample this very ancient crust.

Bringing provisions by float plane to a bush camp on the Canadian Shield, 1978.

Mineral deposits have become the underlying economic basis for communities like Yellowknife NWT, Canada.

Helicopter support can enable rapid and efficient transport in remote regions like those of the far north.

Canoes were the original means of transport for exploration of the Canadian Shield as well as other Precambrian areas of the north, like those of Russia. Not only are they useful on the numerous lakes of the glaciated shield but they can also negotiate streams like this, which interconnect the lakes. Canoes and similar small boats are still an effective method for geologic mapping and prospecting on "The Shield."

Roads, many of them paved, invade the southern portion of the Canadian Shield, which is popular with tourists and sportsmen. This is north of Lake Superior, in western Ontario, in a watershed draining into Hudson's Bay.

Frost polygons. A sure sign of the north, this pattern, developed in glacial sediments, is produced by the freeze-thaw cycle in the presence of permafrost in the far north. Frost polygons like these are also found at high elevations in lower latitudes.

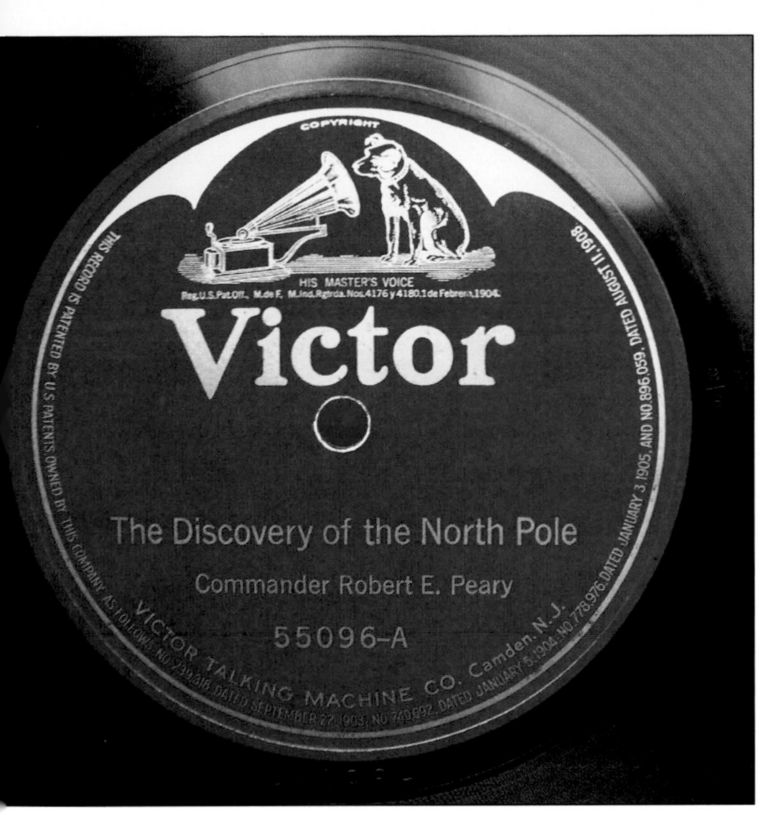

North Pole related recording. The quest for the North and South Poles took place during the early part of the twentieth century. Only after suitable aircraft were developed did the far north (and south) become reasonably accessible. On this recording, Robert E. Peary relates the mechanics of his reaching the North Pole by sledge and dog team in 1909. Old audio recordings relating to historic events can be exciting and valuable finds in jumble and antique shops. (Value range C).

High country. Barren outcrops of Archean gneiss at a 14,000 feet peak in the Colorado Rockies. Massive and hard Precambrian rocks can produce some of the highest portions of mountain ranges.

Archean gneiss at 14,000 feet. Extreme temperature changes and the freeze-thaw cycle at this elevation result in sheets of this hard rock spalling off rock faces. Note the absence of vegetation; the same as found at very high latitudes.

Ancient Rocks and High Elevations

Going up in elevation in a mountain range can be akin to going "up" into high latitudes. Going up in elevation not only mimics "going up" to a high latitude with respect to the presence of high latitude ecosystems, but may also be accompanied by the experience of **going back into deep time with respect to its rocks**. High elevations, like high latitudes, also show evidence of glaciation. Rocks and minerals in areas of high elevations, as in areas of high latitude, show little evidence of weathering, especially chemical weathering. Their rocks appear fresh and "new" looking. The higher one goes, the less weathering that occurs. The reason for this is that high elevations are often regions of extensive uplift. Uplifted areas have had their younger rocks removed by erosion and weathering over geologic time, exposing their core of hard, ancient rocks. Thus, when going into some high mountain ranges, like the Rocky Mountains of Colorado and the Urals of Russia, one can be going backward through time to meet first-hand some of the earliest rock records of the planet.

Glacially carved dome made of Archean gneiss, Longs Peak, Colorado

Talus of gneiss boulders that have spalled from the gneiss outcrops shown in the previous photo.

Different "Earths," a "Broad Brush" Look at the Geologic History of the Earth

Major stages in the evolution of the earth have recently been referred to as **"Black Earth," "Red Earth," "White Earth,"** and **"Green Earth."** Black Earth was the planet four billion (+) years ago with no free oxygen in its atmosphere and only its black, basalt surface, a surface originating from both the cooling of a magma ocean and widespread mafic volcanic activity. **Greenstone belts** of the **Archean** record portions of this earliest of earthly records with its (more commonly black than green) mafic rocks. Gabbros (intrusive mafic rock) and anorthosites (which includes the beautiful labradorite of the Nane Province of the Canadian Shield, Greenland, and Madagascar) are products of this "***Black Earth" stage***. The Black Earth stage includes both the Hadean and the Archean Eras of geologic time. The next, "broad sweep," evolutionary period is that represented by the "***Red Earth.***" The Red Earth stage is that of the early and mid-Proterozoic with its iron formation (BIF), stromatolites, and the beginnings of oxygen accumulation in the atmosphere. The "***White Earth***" stage (also known as Snowball Earth), being a period of alternating coldness (with extensive glaciation) and warmth, includes the late Proterozoic and extends into the Cambrian Period of the Paleozoic Era. The appearance and spread of land vegetation during the Silurian Period of the Paleozoic Era constitutes the last of these "broad brush" evolutionary stages, referred to as the ***Green Earth*** stage, with its photosynthetic vegetation covering the earth's landmasses.

Minerals, Crystals, and "Crystal Consciousness"

During the past twenty years, among a certain group of persons, mineral and natural crystal specimens have taken on an element of mysticism. The author takes note of the fact that one of the appeals of minerals and crystals is aesthetics, a noble and significant virtue. The author, as someone trained in the sciences, at times has met persons who lack this aesthetic appreciation toward minerals, fossils, and other geo-collectibles, which he considers unfortunate. The "crystal consciousness" people certainly have "got it right" with respect to aesthetics and their appreciation of geo-collectibles. It is hoped that these persons, by becoming acquainted with minerals and crystals as geo-collectibles, might also become acquainted with some of the scientific sides of them as well, and perhaps they will even look into fossils and meteorites. With regard toward relating natural objects in a spiritual context, as is done with crystal consciousness, I might invoke the quote by J. B. S. Haldane, the founder of molecular biology, that, **"The Universe is not only stranger than we imagine but is also stranger than we *can imagine*."**

Basalt—***Black Earth Stage***. Cooling of a mafic magma "ocean" some 4.33 billion years ago would have presented nothing but a landscape of barren black rock like this (without the bush to the right). No continents existed at this time yet, but in this environment, during the middle part of the Black Earth Stage, primitive life appeared. This phenomena included photosynthesis, a process which over **millions of years introduced free oxygen into the atmosphere**.

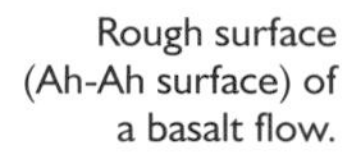
Rough surface (Ah-Ah surface) of a basalt flow.

Value Ranges

Most mineral specimens illustrated here have some sort of monetary value. Some, like native gold or diamonds, can attain values in the six figure range. It might be noted with regard to diamonds that they have not been included in this work although natural diamonds, originating in the earth's mantle, may well have formed during the early Earth period and their inclusion in this work would not be inappropriate.

Sometimes an illustrated specimen will have greater educational or scientific value than monetary, other specimens, like gem crystals, may be the reverse. The author presents monetary value with the understanding that when it conflicts with scientific or educational value, the latter should take precedence.

The value of a mineral specimen is sometimes a property of a specimen's size. The author has also found that smaller (often thumbnail size) specimens make quite attractive photos and, because of their small size, are more affordable to a collector.

A $1,000-$2,000
B $500-$1,000
C $250-$500
D $100-$250
E $50-$100
F $25-$50
G $10-$25
H $1.00-$10.00

Iron Formation—***Red Earth Stage***. This is part of a large iron mine where iron, which was dissolved in sea water as ferris iron, was precipitated to form beds of ferric iron as a consequence of the free oxygen generated from photosynthesis. This free oxygen oxidized the ferrous iron (which was present in great quantities in the oceans) to red ferric iron like that seen in this 2.2 billion year old iron formation exposed in an iron mine in northern Minnesota.

Tillite—***White Earth Stage***. About a billion years ago, large glaciers formed deposits like this (tillite), which are now solid rock. Sometimes the White Earth stage is referred to as the time of "Snowball Earth," as it is believed that it became so cold (at times) that large portions of the oceans were frozen solid.

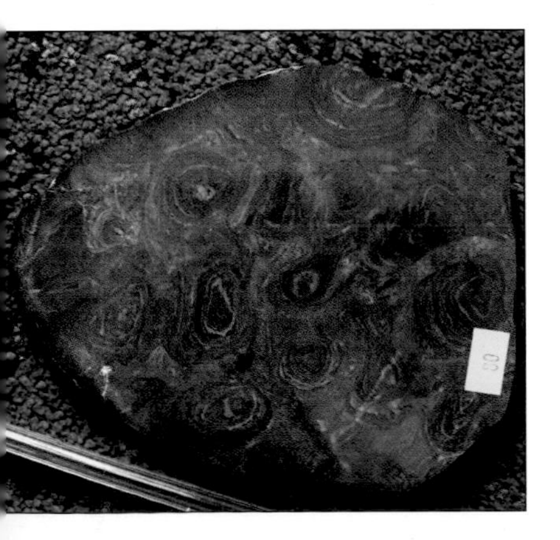

Stromatolite—***Red Earth Stage***. These structures were produced by the primitive photosynthesizing organisms responsible for the Red Earth Period. Stromatolites are fossils produced from the life activities of early life, primarily cyanobacteria—primitive photosynthetic "algae."

Edicaran vendozoans—***White Earth Stage***. Warming after the intense period of late Precambrian glaciation, natural selection and evolution "invented" a peculiar set of large organisms known as vendozoans. Some paleontologists consider these to be a sort of "evolutionary experiment," one that went extinct prior to the Cambrian Period of the Paleozoic Era. Many of these strange fossils appear to be unrelated to anything living today.

Glossary

Cosmic Time. Mega time determined from radiometric age dating on meteorites and moon rocks goes back over five billion years. Astronomical methods of dating stars have determined that some stars, such as those of globular clusters, are as old as ten billion years. Astrophysical methods of age dating have determined, from four degree Kelvin background radiation, that the Universe extends back 13.7 billion years to the time of the "Big Bang." This age dating of the cosmos itself is known as cosmic time.

Geologic Time. Time spans measured from radiometric age dating range from a few thousands of years to over four million years, the age of the earth's oldest rocks (detrital zircons). Geologic time is represented by the geologic time scale and is referenced in terms of phenomena found in the geologic record of the earth.

Gneiss. A high grade metamorphic rock formed from either clastic sedimentary rock like shale or siltstone. Gneiss resembles granite but, unlike granite, it is banded and heterogeneous, often being full of quartz veins.

Hadean. That part of Earth's geologic record (prior to the Archean) that is essentially missing in the earth's crust. It includes the formative stages of the early earth, which includes massive meteor bombardment and a magma ocean.

Intrusive Rock. Igneous rock that has been injected into the earth's crust. This is in contrast to extrusive igneous rock, which was ejected onto the earth's surface.

Regolith. Loose material usually formed from the weathering of underlying rocks. Specifically used here in reference to red clay formed from the weathering of granite and related rocks in a tropical climate.

Uniformitarianism, Principle of. The concept in geology that states, "The present is the key to the past." Developed to explain how geologic phenomena like hills and valleys (topography) formed from running water as seen today, the concept fails, when looking at the long term evolution of the earth. Geologic conditions of the early Earth were different from what they became in later geologic time and the Earth, unlike the Moon (Luna), has **evolved geologically over time**. Uniformitarianism can, however, also refer to the consistency in the fundamental laws of physics and chemistry, which govern the universe. In this context, it is a valid concept even when spanning billions of years.

Bibliography

Hazen, Robert M., 2010. Evolution of Minerals. *Scientific American*, March 2, Vol. 302, No. 3.

Mathez, Edmond A., 2004. A Birthstone for Earth, the oldest terrestrial material in a crystal of zircon, the sometimes diamond substitute that can be a geologist's best friend. *Natural History Magazine*, Vol. 113, No. 4, May 2004.

Fossil scale tree impression (*Lepidodendron*)—***Green Earth Stage***. The mid-Paleozoic (Devonian or Silurian Periods of geologic time) saw the spread of land vegetation over the earth's land areas to produce the Green Earth Period, which we are still in.

Carboniferous Forests—***Green Earth Stage.*** Late Paleozoic land vegetation covered large parts of the northern hemisphere to form the extensive coal beds of the Carboniferous Period of geologic time. This luxuriant vegetation may have (for awhile) raised the earth's atmospheric oxygen level above the twenty-one percent current level.

Chapter Two

Impact of Meteors and the Early Earth

The Oldest Material—Meteorites

Earth formed from the solar nebula (as did the other planets of the solar system and the Sun)! This cloud of primordial material, composed of vast amounts of chondrules and dust, contained most of the matter that would eventually accrete to form the solar system. This nebular material also held short-lived radioisotopes like aluminum-26, which, upon decay, produced substantial amounts of heat. The larger the accreted mass, the more heat that was available. Some of the larger masses, known as planetoids, became hot enough to become incandescent and consequently melt. Melting resulted in metal (primarily iron and nickel like that found in chondritic meteorites) to separate from the molten silicates. Liquid metal is immiscible with liquid silicates and thus separates from it. That is what the pallasite meteorites are all about. With the earth, given its considerable mass (and hence considerable gravity), this process produced a metallic core composed of the nickel-iron of chondritic material. Overlying this dense core was immiscible silicate material forming a thick shell surrounding the core. This mantle over geologic time separated (fractionated) into a predominantly denser fraction (mafic and ultramafic silicates) and a lighter fraction, the felsic silicates. Over long spans of time these fractions separated and then the felsic silicates "floated" upon the denser mafic and ultra mafic material. These first felsic silicates formed the beginnings of the continents, the proto-continents. The metallic core that formed would concentrate not only most of the earth's nickel, but also other siderophile elements like chromium, cobalt, and the platinum group metals like palladium and iridium, as well as platinum.

The round objects seen scattered through this slab of a stony meteorite are known as chondrules. They represent the most primitive matter that can actually be examined and studied. They are the "original stuff" from which the solar system was made. The chondrules in chondrites formed before the planets and the solar nebula formed. This was from some sort of condensate of gas given off by either a nova or a "dying" red giant star, all of which predated the formation of the solar system. (Value range E).

Chondritic meteorite. A chondrite made up of fragments of older, broken-up meteoroids. One of the predominant processes in the early solar system was **impact**. These impacts took place either between two meteoroids or a meteoroid and an early formed planetoid. The majority of material in this stony meteorite is composed of silicates, the grey objects are metallic nickel-iron, a component of most meteorites. (Value range F).

Chondrules compose most of this NWA (Northwest Africa) meteorite. This is another example of a very primitive form of extra-terrestrial matter. Large numbers of meteorites have been found and collected recently in the Sahara Desert and these have made primitive meteorites (like this) much more available and accessible. (Value range G).

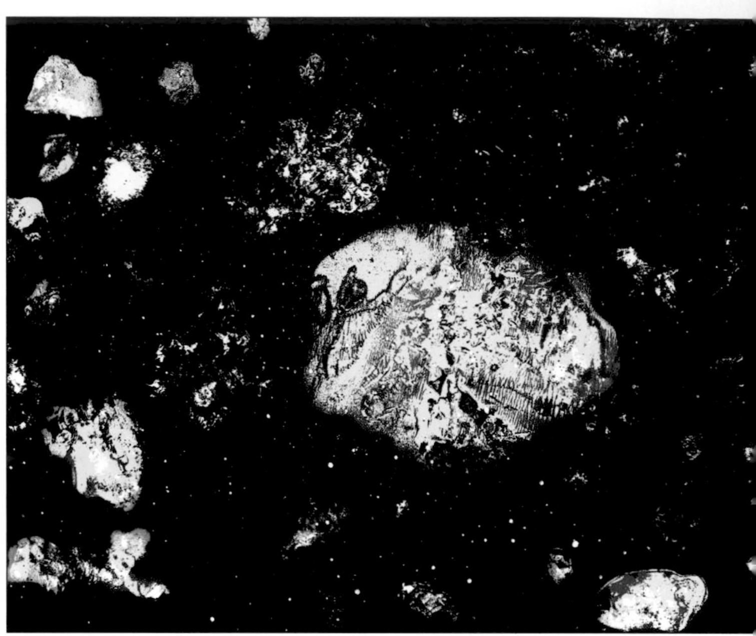

Incandescent masses of chondritic material in space 5.5 billion years ago. Nebular material condensed to form masses of hot molten matter like this. The matter is incandescent as it contains a lot of radioactive material which, upon decay, produces heat. Hot masses of incandescent material like this, upon impact, clumped together and grew as does a large snowball composed of wet snow. The metal, which was in the chondritic material, upon liquefaction, consequently separated from the silicate as these two materials are immiscible.

More chondrules in a primitive meteorite (Allende) which fell in Mexico in 1969. (Value range G).

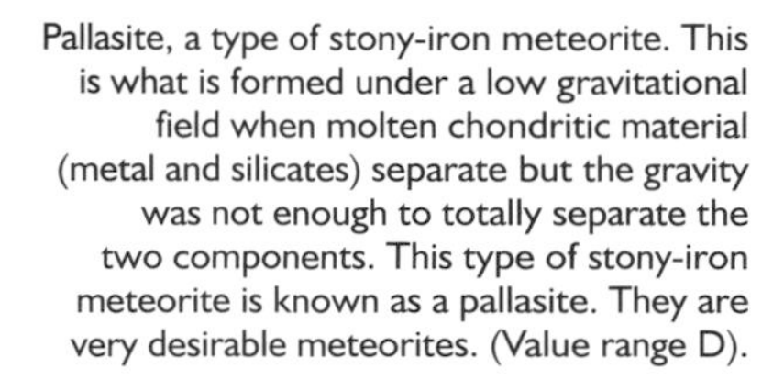

Pallasite, a type of stony-iron meteorite. This is what is formed under a low gravitational field when molten chondritic material (metal and silicates) separate but the gravity was not enough to totally separate the two components. This type of stony-iron meteorite is known as a pallasite. They are very desirable meteorites. (Value range D).

Nickel iron (metallic) meteorite. This is what forms when metal concentrates in the center (core) of a mass large enough to have a significant gravitational field. The pattern (Widmanstatten Figures) of interlocking crystals formed from the slow cooling of the nickel-iron alloy. When sufficient amounts of chondritic material "accreted" or clumped together, the heat of radioactivity melted the entire mass, forming a planetoid. Metallic iron in the accreted mass formed a center or core. This core was covered by a layer of lighter, immiscible silicate material to produce a mantle. When the planetoid cooled, it was impacted by another large object and fragmented. This metallic meteorite, with its interlocking crystals, is a portion of the crystallized metallic core of the planetoid. **Metallic meteorites derived in this manner are known as siderites.** (Value range E).

Silicate material formed from the mantle of a mass large enough to melt. Both this silicate meteorite (achondrite) and the previously shown metallic meteorite are secondary types of meteorites or derived meteorites. Being secondary to the primary chondritic types, they are less primitive than are chondritic meteorites. (Value range E).

Fragments of another type of achondritic meteorite. Secondary silicate material ultimately derived from the melting of chondritic meteoritic material. The light color of this achondritic meteorite indicates a low iron content of the mantle of the parent planetoid or asteroid. (Value range E for group).

Meteor Crater, Arizona. This is a very young crater formed on the surface of northern Arizona about 45 thousand years ago. Impact craters like this don't survive very well on the earth's surface because processes like weathering and erosion remove them. Long term geologic processes, such as sedimentation (covered by sedimentary rock) and subduction, also have removed or buried most earthly evidence of impact. Impacts like those shown here, which are easily recognizable, are for the most part geologically young.

Twenty-five million year old impact site (astroblem). This circular structure in southern Germany is a sediment filled crater formed when a large meteorite (or small asteroid) impacted during recent geologic time (Late Cenozoic).

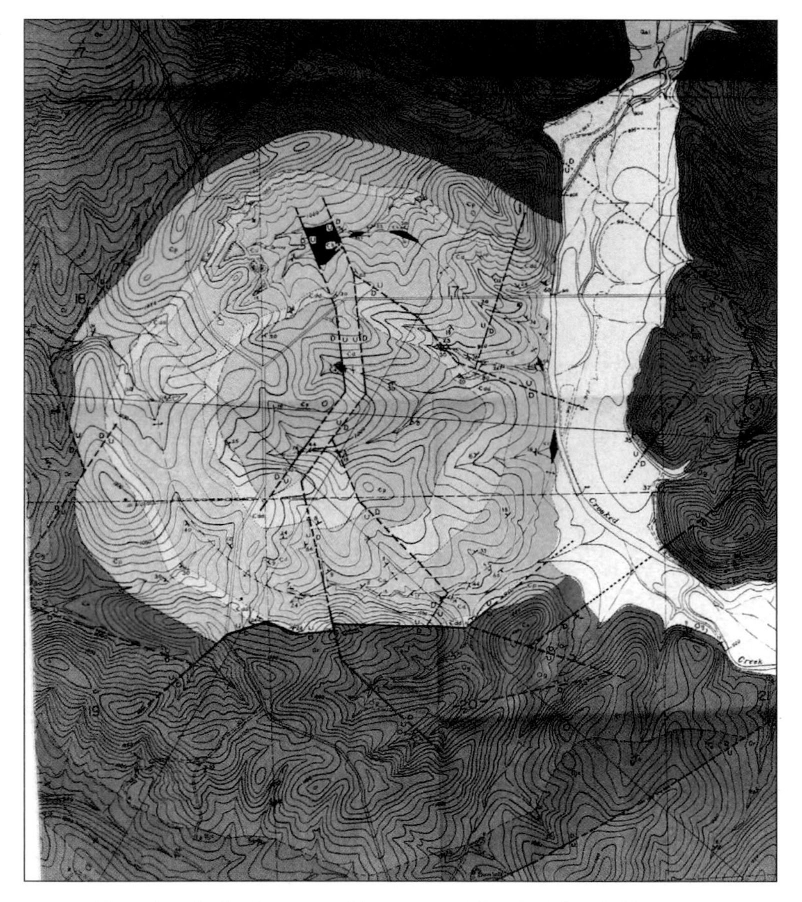

Three hundred million year old impact site (Crooked Creek, Missouri, astroblem). The impact producing this structure took place considerably earlier than did those forming the previously shown structures. **However, the Crooked Creek astroblem is *still* geologically young**.

Impact breccia from the Crooked Creek astroblem. High velocity impact breaks up material into fragments known as clasts. These fragments are then later cemented together with iron oxide to make a breccia.

Five hundred million year old snail impressions. These were damaged by an impact from 300 million years ago, the impact that formed the Crooked Creek astroblem. (Value range F).

Another 500+ million year old fossil affected by impact at the Crooked Creek astroblem. This snail-like fossil (*Scaevogyra*) is Cambrian in age. Cambrian age rocks were brought up by rebound at the ground zero portion of the structure, otherwise they would not be at the earth's surface but would rather be still buried under younger rocks. (Value range G).

Early Processes on the Earth

Geological evolution and geochemical reactions have been sorting out elements of the earth since it was formed, separating elements through geological processes such as cooling and crystallization, magmatic separation and volcanism. The geologically significant processes of atmospheric and hydrospheric weathering and sediment generation also have been significant on the earth. In other words, fundamental geologic processes that do not take place on asteroids were involved on Earth, and have only barely taken place on Mars and the Moon.

Some of the earth's oldest rocks, however, retain some of the primitive characteristics and signatures of meteorites. These signatures will be discussed below, with discussions of basic geology related to some of the planet's oldest terrain.

Basalt. This black, relatively heavy rock formed the surface of the early earth after the surface cooled. (Sometimes basalt can be light, as when it contains many gas bubbles). This geologically recent lava flow represents a phenomena that has existed since the earth formed. Basalt represents the remnants of silicate material that ultimately came from chondritic material. Basalt like this has been extruded from the earth's interior for over four billion years. This scene (except for the land plants) could have been from any time over the past 4.5 billion years.

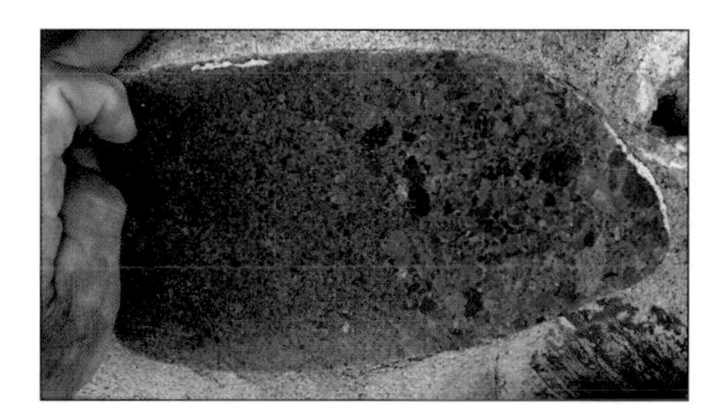

Impact breccia? From 3.2 billion year old rock from northern Minnesota. Breccias would appear to be common in the geologic record of the early earth but they are not. This may be because so much of this early record has either been destroyed or has been modified beyond recognition by metamorphism. Note that the breccia is dark colored. It is composed of mostly mafic material, a type of rock composition typical of the early earth.

Basalt actually is a rather dense rock. It, however, can be full of gas bubbles as are these basalt boulders; this makes them quite light. In Archean basalts such holes are always filled in by minerals so that these ancient rocks are anything but light.

Geyser and hot spring vent, Yellowstone Park, Wyoming. Geothermal activity today is a relatively uncommon phenomena. However, these geothermic features are one of the major attractions of Yellowstone. In the early earth, such geothermal phenomena is presumed to have been common. However, as is the case with evidence for impact, the early earthly geologic record is sparse in its documentation of these phenomena. Presumably this is because much of the evidence for such activity has been destroyed.

Three billion year old basalt (right) intruded by younger granite (reddish). The white material in the basalt is quartz, which was emplaced later, probably when the granite was intruded. West central Manitoba.

Close-up of Archean basalt on the Canadian Shield. Basalt like this (along with ancient black sedimentary rock known as greywacke) is some of the oldest rock of the earth's crust. The circular structures in the center of the photo are lava "pillows" in pillow lava; basalt which was extruded on the ocean floor.

Archean granite on the surface. Granite is formed from high silica magma (magma that separates from basalt in the mantle) being added at the base of the earth's crust by a process known as underplating. To expose such deep seated rock on the surface takes at least hundreds of million of years; that is why most granites are geologically old.

Granite is a typical continental rock. It represents the light fraction of silicates that separated from much more common mafic material composed of iron, magnesium, and calcium alumina silicates. Granite is formed from masses of molten felsic rock (magma) rising and adding itself to the underside of a thin crust, felsitic underplating. This process builds the crust from below, thereby adding to and thickening continental crust. Only over hundreds of millions of years can this granite be exposed on the earth's surface after overlying rocks were weathered away.

Nickel and Chromium Deposits of the Early Crust

Localized concentrations of nickel minerals occur in some of the most ancient rocks of the earth's crust when the metal-silicate separating mechanism wasn't totally engaged. These nickel deposits, linking back to some of the last stages of massive bombardment of the earth by asteroids can today be the source of valuable nickel deposits like those of Sudbury, Ontario, and Flin Flon, Manitoba—both in Archean rocks of the Canadian Shield. It is a reasonable hypothesis that these ancient nickel deposits formed from asteroids absorbed into the early crust some three billion years ago. The crust was thinner, and less rigid at that time than it is today. An asteroid hitting the early earth not only produced large craters in the early felsic crust, but also may have been absorbed into the crust. If this happened, the nickel was remobilized and combined with sulfur to produce nickel-iron sulfides mined today in very old portions of the earth's crust.

Seemingly, asteroids impacting planetary surfaces of the inner solar system three billion years ago formed not only earthly nickel deposits but also impact sites, one of which may be the circular configuration at the southern end of Hudson's Bay. Other planetary bodies of the solar system, like the Moon, were also impacted by similar large asteroids, which released molten mafic material from below the Moon's surface. Thus, from the outpouring of black, molten lava, triggered by these large impacts, was the maria of the Moon formed. Maria on the lunar surface has a much lower crater density than does the massively cratered lunar uplands, which formed over 4.5 billion years ago, during the most intense period of bombardment. A similar asteroid impact on Mars, about the same time, may have initiated the volcanism responsible for the huge volcano on Mars known as Olympus Mons.

Many of the earth's oldest rocks are found in geologic regions known as shields. These are usually relatively low lying regions, but depending upon latitude and more recent uplift, can still be rugged land, often with lots of hard rock outcroppings. The major shields of the world are the Brazilian Shield, large parts of western and northern Australia, the Finno-scandanavian Shield, Ankara Shield, and the largest, the Canadian Shield of North America. The cores of some of the world's mountain ranges may also consist of ancient rocks. Mountains like the Rockies, the Urals, the Appalachians, and the Finders Range of Australia have such cores. Smaller, shield-like regions such as the massifs of Europe, the Llano Uplift of Texas, the Ozark Uplift of Missouri, and the Black Hills of South Dakota have also seen ancient rocks pushed up to the earth's surface, where they then became accessible.

Flin Flon, Manitoba. Extensive nickel mining occurs in ancient rocks of the Archean of central Manitoba, Canada. Smelting and mining of nickel from these early deposits form the economic base of this region. The nickel mined here appears to have originated from meteoroids or asteroids impacting the crust and, because it was hot and mobile, became incorporated into it along with its nickel content.

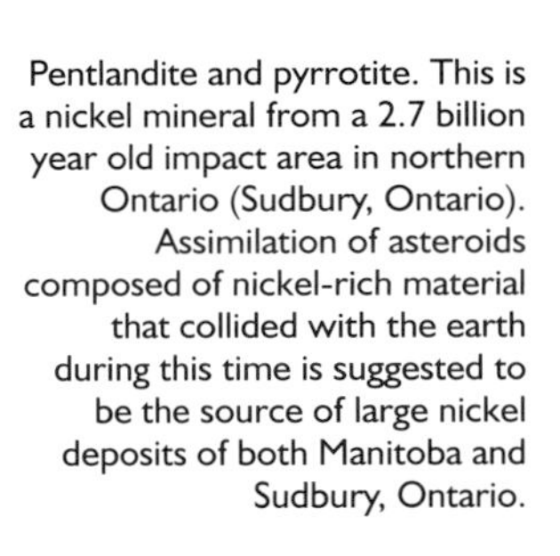

Pentlandite and pyrrotite. This is a nickel mineral from a 2.7 billion year old impact area in northern Ontario (Sudbury, Ontario). Assimilation of asteroids composed of nickel-rich material that collided with the earth during this time is suggested to be the source of large nickel deposits of both Manitoba and Sudbury, Ontario.

Nickel smelter at Flin Flon, Manitoba. Archean rocks of central Manitoba, on the Canadian Shield, have numerous nickel deposits. Some of these contain large amounts of nickel minerals, hypothesized by some geologists to have originated from impact with the early earth.

Millerite. This mineral, composed of nickel sulfide, is believed by many geologists to have been derived from meteorite or asteroidal material, which bombarded the earth during the early period of **heavy meteoritic bombardment**.

Flon Flon, Manitoba. Ancient Archean rocks outcrop in the vicinity of this small city, which is based upon nickel mining and smelting. Rocks in the area show up quite nicely as they have been stripped of muskeg to cut down on the mosquito population—insects which exist in vast numbers on the Canadian Shield and elsewhere in high latitude regions of ancient glaciated rocks like the Kola Peninsula of Russia.

Zaratite. This is a secondary nickel mineral from very ancient rocks in Tasmania. Rock formed during the early part of the earth's history often contained nickel minerals, nickel that was associated with the terminal stages of heavy meteoroid bombardment. (Value range G).

Millerite, Missouri. This spectacular spray of millerite needles (nickel sulfide) may have originated from the weathering and re-precipitation of nickel, which originated from a (now gone) strewn field of meteorites. *Courtesy of Glenn Williams*

Chrysoprase, Australia-polished pebbles. This attractive, semi-precious stone is often made into "rockhound jewelry." The green color comes from nickel, nickel that may have come from meteoritic material associated with this early period of Earth history—the period of heavy meteoroid bombardment. Large portions of Australia expose ancient rocks that sometimes, as with chrysoprase, can contain nickel minerals (or compounds). (Value range G, single specimen).

Chromite mine at high elevation, Stillwater complex, Montana. Large deposits of chromite (chromium oxide) occur in the Stillwater Complex, an area of diverse and unusual Archean igneous rocks in Montana's Absaroka Range. Chromium like nickel is associated with metallic meteorites and (as is the case with nickel) may have come from the last stage of meteorite bombardment associated with the early earth.

Chromite from the Stillwater Complex. This is typical high grade chromium ore (chromium oxide) from the mine in Montana's Stillwater Complex.

Chromite, Stillwater Complex, Montana.

The Archean

The most ancient rocks of the earth were formed during the earliest (and least understood) portion of Earth's history, the earliest part of the geologic time scale, known as the Archean Era. Archean rocks range in age from four billion years (the oldest currently known) to 2.5 billion years (2.5 Gyr—giga or billion years). This portion of geologic time is defined as that part of Earth's history when the continents either did not exist or they were so small that geologic processes prevailed that were somewhat different from those of later geologic time.

The Canadian Shield

This is one of the largest areas of ancient rocks exposed on the surface of Earth. The Canadian Shield is divided into a number of sub-provinces, of which the Nane and the Great Slave Lake provinces are the oldest. Almost as old as these two provinces is the Superior Province, a part of the shield that exposes Archean rocks generally from 3.2 to 2.5 billion years old. As is the case with other ancient shield areas, the oldest rocks of the Superior Province consist of greenstone belts in which metasedimentary rocks are found in a vertical position, usually surrounded by gneiss or granite gneiss. A few of the beds in the greenstone belts contain inter-bedded limestone carrying evidence of early life in the form of stromatolites. Some of the black slates of the ancient greenstone belts can also contain considerable amounts of carbon that may be of biogenic origin. The prevalence of carbon in these black, slaty rocks has been explained as being a residue formed by inorganic carbon-containing-compounds known as a coacervate or from organic material derived from early life like chemosynthetic or photosynthetic bacteria.

Gold occurrences, sometimes in quartz like this, can be characteristic of Archean rocks and terrains.

Black, carbon-rich slate, Black Hills, South Dakota. This black slate was originally black, deep sea mud. Organic material was probably the source of the carbon, which gives this slate its deep black appearance. Archean rocks like this can contain large amounts of carbon, which may have come from vast amounts of carbon dioxide sequestered during early planetary evolution. The major component of the atmosphere at that time was carbon dioxide, which was utilized by early photosynthetic organisms, possibly photosynthetic bacteria, to produce biomass. The greenhouse effect of the carbon dioxide atmosphere may also have counterbalanced the lower energy output of the Sun at that time.

Black slate outcrop in the old (above ground) portion of the Homestake Mine, Black Hills, South Dakota. Black slates of Archean age in the Black Hills (and elsewhere) can be auriferous (gold bearing). The tectonics of the early earth appear to have been conducive for emplacement of gold as many (most) of the earth's large gold occurrences are found in Archean rocks, often associated with rocks of greenstone belts.

The gold that formed Colorado's mid-nineteenth century gold rush was (in part) associated with pyrite in quartz veins of Archean age.

Pyrite in greenstone, Timmins, Ontario. Archean rocks in western Ontario and northern Ontario, Canada, on the Canadian Shield, are source rocks for large scale gold mining and production. This is auriferous (gold bearing) meta-mudstone from Timmins, Ontario, a major source of gold production on the Canadian Shield. Rarely is visible gold seen in these rocks; rather gold is disseminated in the rock in very small specks or is incorporated into pyrite crystals, as seen here.

Auriferous pyrite crystals in Archean greenstone, Hemlo Mine, western Ontario.

Yellowknife NWT, Canada, 1958. Archean rocks at the northwestern edge of the Canadian Shield produce about one third of Canada's gold production. Gold mining is centered around Yellowknife, the major town in Canada's Northwest Territories.

Zircon crystal, Kipawa River, Quebec. Zircon is a mineral of the relatively uncommon element zirconium. Chemically, it is a zirconium silicate. Zircon crystals of large size, like this, are characteristic of Archean terrains. Zircon usually contains some uranium, where the element is tightly held within the minerals crystalline lattice. Because of this, zircons are especially valuable for determining radiometric age dates of the rocks in which they occur. The oldest age dates for any terrestrial mineral or rock has been obtained from zircons found in early Archean conglomerate from Australia. These zircons give a date of 4.2 billion years. This is the only age date of a terrestrial mineral or rock from the Hadean Era.

Glossary

Asteroid. A "small" object (1 km-100 km in diameter) that orbits today and exists today.

Chondritic Meteorites. These are the most primitive of meteorites and are also the most primitive forms of matter that one can actually handle. They are characterized as containing small spherical objects known as chondrules.

Chondrules. Small (3-10mm) spherical objects formed in the solar nebula prior to the formation of the solar system. Chondrules at one time were molten from their short-lived radio-isotope content which produced considerable amounts of heat.

Meteoroid. A chunk of extraterrestrial material a few centimeters to over one kilometer in diameter. Objects larger then this would be asteroids. No definite size determines when a meteoroid becomes an asteroid or a planetoid.

Planetoid. An extraterrestrial object that existed during the early stages of the solar system and accreted from hot, slushy chondritic material. Planetoids and asteroids are similar in size, but separated by time. Planetoids formed directly from chondritic material during the early history of the solar system. They were formed of material high in radioactive isotopes so that they were hot, sometimes even molten. An asteroid is a planetoid size body that exists today (without the radioactive material).

Siderophile Elements. Elements attracted to iron when a mass is in a molten condition. These elements include nickel, cobalt, chromium, the platinum group elements, and sulfur.

Ultramafic Silicates. Igneous rock made primarily of silicate minerals like olivine, and various types of pyroxene. Ultramafic material is what is believed to compose most of the earth's mantle. It also is not too far from the silicate material found in chondritic meteorites.

Visible gold in quartz. Most areas where Archean gold is found rarely exhibit the gold as visible crystals like this. The gold, instead, is highly disseminated through greenstone or slate-like rocks. Why Archean rocks are sometimes gold bearing is not completely understood (as is the case with a lot of Archean geology). (Value range E).

Zircon crystal. This well formed (euhedral) zircon crystal is considerably younger than that of the previous photo. It comes from Paleozoic intrusive igneous rock in Mont-Saint-Hilaire, southern Quebec. (Value range F).

Chapter Three

Pegmatites and Pegmatite Minerals —Ancient Action Underneath the Crust!

Granite is a rock especially characteristic of Earth's continents. Generally it is formed when felsic magma is added to continental crust from below, a process known as mantle underplating. Granite can accumulate from underplating over spans of geologic time to form great masses known as batholiths. These masses of granite, being resistant to weathering and erosion can sometimes form whole mountain ranges. They can do this if they were uplifted fast enough and whatever was above them was removed by erosion. Pegmatites, the source of many beautiful and rare minerals, represent small parts of granite masses that cooled especially slow, so that rare elements in the magma were concentrated as a residue in the slowly cooling mass. Elements that do not readily fit into the crystal structure of common components of granite, like feldspar and amphibole, become concentrated in the residue, which becomes a pegmatite. In another view, pegmatites represent the last remnants to cool and crystallize in a granitic magma—remnants which often carry a package of elements that don't fit into the crystalline lattices of common silicate minerals. **Pegmatites are especially characteristic of ancient terrains of the earth, as they are associated with granite and it usually takes long intervals of geologic time to expose granite at the earth's surface.**

Pegmatites, Stars, and Gems

Pegmatites usually are associated with granite, an intrusive rock almost unique to the earth. Most of the crystals composing pegmatites are the same minerals found in normal granite, only they are larger. Some pegmatite minerals, however, can also be uncommon ones and some pegmatites can be full of minerals made up of rare elements. Such pegmatites, known as rare earth pegmatites, can be the source of these rare elements. Rare minerals in pegmatites can also include many very hard minerals, of which many are gem minerals. Three light elements, lithium (Li), Beryllium (Be), and Boron (B) are especially characteristic of many pegmatite minerals. These low-atomic-number elements are rare in the universe, especially when compared to hydrogen and helium on one end (of the periodic table) and carbon, nitrogen, fluorine, and argon on their right. Lithium, beryllium, and boron are rare because fusion processes in stars did not directly produce them—what little there was being rare and widely diffused in both primitive meteoritic material and in the mafic magma and rock derived from it. The process of granite formation concentrates these elements in the residue which goes to form pegmatites. This is because Li, Be, and boron do not fit into the space lattice of normal silicates like feldspar or mica. When these rare elements become concentrated in the pegmatite "residue," they become responsible for most of the gem silicates, like tourmaline (contains beryllium), spodumene, kunzite and beryl (contains lithium), and topaz (contains boron).

Granite and pegmatite outcrops, Black Hills, South Dakota.

Pegmatite (and granite intrusives) "peeking out" from vegetal cover in the Black Hills of South Dakota. Because pegmatites (and granite) produce so little soil on weathering in a temperate climate, they become obvious and stick out. The rock into which pegmatites (and granite) were intruded is less resistant to weathering than are pegmatites with their quartz. The rock into which the pegmatites are intruded are softer and thus more readily form soil, which therefore supports vegetation. Note how the pegmatites are more or less evenly dispersed over this area of 2.5 billion year old rock. It takes long spans of geologic time for weathering and erosion to expose pegmatites like these on the earth's surface.

Conspicuous pegmatite outcrops, Black Hills, South Dakota.

Close-up of "rose" quartz outcrop, Black Hills, S. Dakota.

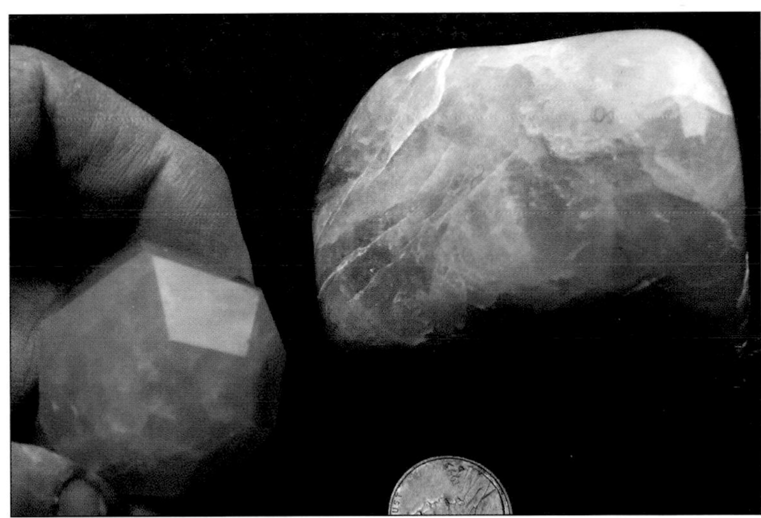

Rose quartz. Outcrops of rose quartz like that shown previously fade upon exposure to weathering and the atmosphere. These are two pieces from fresh pegmatite exposures that retain their pronounced pink color. Pegmatites are composed of especially large crystals, which formed from slow cooling of felsic magma, often containing a lot of water and lighter elements. (Value range F, both specimens).

Massive quartz (rose quartz) outcropping in the Black Hills. Quartz masses often make up major portions of pegmatites as they do in this outcrop. Outcrops of rose quartz lose their pinkish color when exposed to weathering and erosion. Quartz is hard and withstands weathering, however over spans of geologic time it is also removed by weathering.

Mica "book" from previously shown Black Hills outcrop. This mica crystal was collected from just above the massive bed of quartz.

Black tourmaline (schorl) from the area of the quartz outcrop shown in the previous photo. It was associated with mica "books."

Black tourmaline embedded in quartz from pegmatite in the Medicine Bow Mountains, Wyoming. The grey gneiss into which this pegmatite was intruded comes from some of the oldest rocks in the Rocky Mountains. Some of these rocks are over 3 billion years old, dating back to the Mid-Archean. (Value range F).

Orthoclase-rich pegmatite dike near Mt. Zirkel, Colorado. This pegmatite dike has been intruded (injected into dark grey Archean meta-greywacke). Pegmatite dikes and masses intrude into these ancient Archean rocks, which form part of the Wyoming Crustal Province, a very ancient part of the Rocky Mountain region of northern Colorado, Wyoming, and southwestern Montana.

Massive pegmatite outcrop near Espanola, Ontario. The sparkling crystals are muscovite mica.

Pegmatite chunk from the same region shown in previous photo (Mt. Zirkel, Colorado). The pink mineral is orthoclase, the white is quartz, and grey is mica (muscovite). This is one of the most common types of pegmatite found in Precambrian rocks.

Pegmatite chunk from Archean basement rocks of northwestern Quebec. The rock sequence from which this was collected is a candidate for some of the earths oldest rocks. Gneiss, which was intruded by this pegmatite, crops out along the east shore of Hudson's Bay. This gneiss recently (2010) has been dated at over 4.0 billion years (>4.0Gy), making it a candidate for the **earth's oldest rock**.

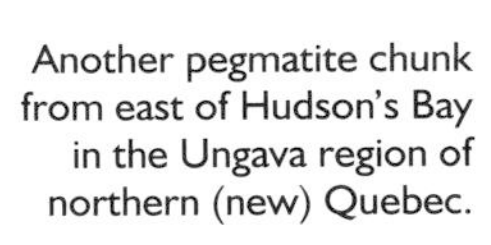

Another pegmatite chunk from east of Hudson's Bay in the Ungava region of northern (new) Quebec.

Pegmatite outcrop in the Black Hills near Keystone, South Dakota. The black mineral at the bottom left is black tourmaline (schorl).

Graphic granite. Dixie looks intently at the peculiar pattern shown in this granite slab. This pattern, referred to as graphic granite, is associated with granite found adjacent to pegmatites. It is sometimes used for counter tops because of its interesting pattern.

Graphic granite hearts! The interesting pattern in graphic granite (which resembles writing) sometimes accentuates a carving. These graphic granite hearts are a somewhat "hokey" example. (Value range G).

Archean pegmatite, western Montana.

Pegmatite made up of mica (Muscovite) "books," quartz (grey), and tourmaline (black). Keystone, Black Hills, South Dakota.

Aerial view of Archean intrusive rock (granite), Almeyer Lake, NWT Canada.

Pegmatite dike, western Ontario, exposed along the Trans-Canada Highway. This pegmatite has been intruded into Archean gneiss. It is a typical outcrop of a pegmatite dike associated with high grade metamorphic rock, which often is of Archean age.

Mica crystals from pegmatite near Val-d'Or Quebec, Canada.

Close-up of the previously shown pegmatite dike.

Cluster of mica crystals from Crystal Peak, Teller Co., Colorado.

Orthoclase crystal, Crystal Peak, Teller Co., Colorado. The Crystal Peak pegmatites near Florissant, Colorado, are part of the Pikes Peak (granite) batholith and have supplied attractive pegmatite minerals for over 100 years. Especially nice are the blue-green microcline (amazonite) crystals that occur in abundance, often interspersed with orthoclase crystals like this.

Granite, Granitization, and Pegmatites

Granite can form either from magma that rose from deep beneath the earth's surface (and in the process adding itself to the crust as continental crust) or from a process known as granitization. Granitization is a process where sedimentary rock, like shale, subjected to deep burial, becomes converted to either granite or to the banded and often ancient metamorphic rock known as gneiss. With either of these cases, quartz-rich fluids that contained considerable amounts of water accompanied the granite or gneiss formation, and these fluids, upon slowly cooling, crystallized to form pegmatites.

Single amazonite crystal, Crystal Peak, Colorado. These blue-green microcline crystals have been widely distributed among collectors, museums, and dealers.

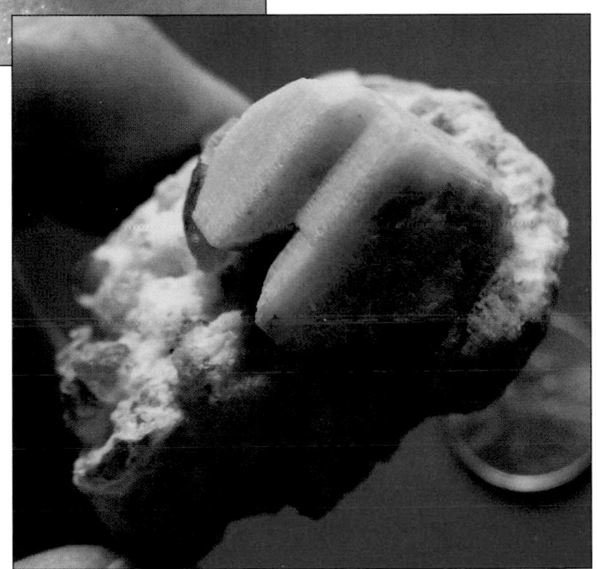

Microcline in quartz, Crystal Peak, Colorado.

Microcline on (somewhat) ugly quartz. Crystal Peak, Colorado.

Rich blue-green microcline with somewhat nasty-looking quartz from Crystal Peak, Colorado.

Amazonite, Crystal Peak Pegmatite. Mineral specimens from this locality have been widely distributed among mineral collectors and rockhounds for over 100 years.

Group of microcline crystals from an old collection. Mineral collections amassed over the past 100 years usually will include specimens of this blue microcline (amazonite or amazonstone) from Crystal Peak, Colorado.

Very ancient amazonite crystal (blue-green microcline) from the Kola Peninsula, Russia. The **early Archean rocks** of the Kola Peninsula of northern Russia extend above the Arctic Circle. These have been intruded by pegmatites, some of which are the source of many rare minerals. Crystals of blue-green microcline are one of the more frequently seen minerals from these pegmatites. (Value range F).

Stone Mountain, Georgia. The monolith known as Stone Mountain is composed of grey granite that contains dikes of tourmaline bearing pegmatite. Carvings on the face of this mountain (which is near Atlanta, Georgia) make it the Confederate Mount Rushmore. Stone Mountain bears the figures of Robert E. Lee, Stonewall Jackson, and Jefferson Davis of the 1860s southern Confederacy.

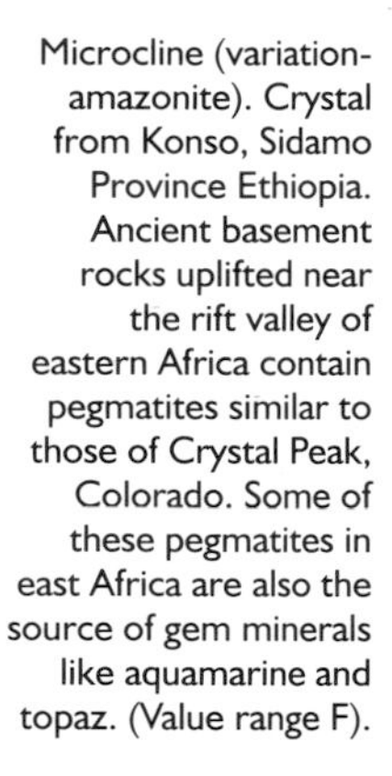

Microcline (variation-amazonite). Crystal from Konso, Sidamo Province Ethiopia. Ancient basement rocks uplifted near the rift valley of eastern Africa contain pegmatites similar to those of Crystal Peak, Colorado. Some of these pegmatites in east Africa are also the source of gem minerals like aquamarine and topaz. (Value range F).

Close-up of Stone Mountain bas-relief carving.

Pegmatite vein intruded into metamorphosed rocks of the Mid-Proterozoic Belt Series, central Idaho. The Belt Series is a sequence of 1.5 billion year old rocks, which to the east in Montana were little affected by metamorphism and carry a variety of fossils, especially well preserved stromatolites (see chapter six). These rocks of the Belt Series in Idaho were metamorphosed and preserved from erosion by being deeply buried in the earth's crust as a part of the Cordilleran Geosyncline. Their deep burial allowed for intrusion into them of thick "veins" or pegmatite dikes like the one shown here.

Stone Mountain Memorial. Efforts to establish this "Mount Rushmore of the South" began in earnest after the Great War (WWI). This song, quite popular in the South during the 1920s, contributed to the cause. The performing artist, Vernon Dalhart, was a popular "Country and Western" singer before the genre was named as such. Dalhart's recordings, popular with rural America at the time, include of an interesting bit of Americana—now little known except to 78 RPM record collectors.

"Oh, Stone Mountain, Oh-Stone Mountain,—preserve thy story well and to our children's children the story you can tell"!

TO INVESTORS.
A Valuable Mica Property
FOR SALE.
Address JOHN C. MERRILL, Attorney-at-Law, Easton, Pa.

Turn of the twentieth century ad for a pegmatite property in the southern Appalachians. Mica from pegmatites was widely used in a variety of ways, which included stove windows, carriage windows (Ising-glass), and two decades after the ad was published in older electronics for the support of filaments in vacuum tubes (De Forrest Audions).

Tourmaline crystals, Stone Mountain, Georgia. Groups of black tourmaline crystals (schorl) occur in dikes shot through the grey granite of Stone Mountain, Georgia.

Another use for thin, cleaved mica crystals from pegmatites was for diaphragms in acoustical audio devices like stethoscopes and phonographs. Its hardness coupled with its flexibility lent it to use in acoustic phonograph diaphragms as it emphasized the higher frequencies of sound in both playback and recording.

Autunite flakes on feldspar, Stone Mountain. Excavations in the mid-1980s yielded flakes of this secondary uranium mineral on feldspar. Autunite is fluorescent under a black light, as are many other secondary uranium minerals. Uranium minerals occur in other Appalachian Mountain pegmatites like those of Spruce Mountain, North Carolina. (Value range G).

Another mica diaphragm, this time in a French (*Pathé*) acoustical phonograph. Use of natural materials for new technology was commonplace in the nineteenth and early twentieth centuries. The range of manmade materials was, at the time, considerably limited. Thomas Edison in his invention factory at Menlo Park, New Jersey, had thousands of different natural materials (including a large selection of minerals) to utilize as needed in his inventions.

Quartz masses in the Ouachita Mountains of Arkansas. Large numbers of quartz crystals and quartz masses like this occur in central Arkansas. Some indications suggest that this quartz is the uppermost portion of what are still buried pegmatite bodies. Because they are only Paleozoic in age (Early Paleozoic, approximately 450 Gy), their underlying pegmatites have not been exposed by erosion, as have pegmatites of greater age in Precambrian rocks characteristic of geologically older terrains.

Group of medium size quartz crystals from Arkansas.

Rubellite (pink tourmaline), from near Pala District, San Diego Co., California. Tourmaline is a complex lithium containing silicate mineral. Rubellite occurs associated with lithium bearing pegmatites, pegmatites formed from water bearing magma, which contained an excess of this rare, light element. (Value range F).

Quartz and feldspar crystals, northern Georgia. Quartz crystals similar to those found in Arkansas occur with feldspar (sodium plagioclase) crystals in Georgia. The occurrences are somewhat similar to those of Arkansas, but they are geologically older and therefore more uplift and erosion has been able to have taken place, allowing deeper exposure into the pegmatite mass, deeper than is the case with similar but younger quartz masses of Arkansas.

Blue tourmaline, Ingersoll Mine, Keystone, South Dakota. Blue beryl is uncommon and is found only at a few localities worldwide, which includes the Black Hills. The presence of different amounts of lithium, beryllium, and boron imparts the color to this attractive mineral. So long as iron is absent, these delicate colors can occur. When iron is present, it colors the tourmaline black or dark green.

Rubellite (pink tourmaline) crystals in lepidolite (lithium mica) from Pala District, San Diego Co., California. Lithium, beryllium, and boron rich pegmatites produce spectacular and beautiful minerals, many of which are gem minerals (like rubellite, if it is fracture free, which is rare). Pegmatites, rich in these three elements, but low in iron, are less common than are those which carry black tourmaline, the more common iron bearing tourmaline known as schorl. The elements lithium, beryllium, and boron are themselves unique and peculiar in their being the **only low atomic weight elements** that are **not common components of stars**. They appear to have formed indirectly from fusion reactions taking place early in the history of the Milky Way Galaxy. They may have formed even during the early history of the Universe. Usually relatively rare, they are concentrated in lithium pegmatites to form odd and attractive minerals like this. (The Pala district pegmatites are geologically young, Mid-Mesozoic.) (Value range F).

Blue tourmaline, Keystone, South Dakota. Another specimen of blue tourmaline from the Black Hills. This specimen was collected in the 1940s and came from the collection of geologist Joe Schraut. Mr. Schraut was one of the founding members of the St. Louis Mineral and Gem Society. (Value range F).

Rare Earth Pegmatites

As pegmatites represent the last stages in the cooling of large quartz-rich masses of magma, they can form minerals that normally are not found associated with granite. Normal pegmatites can contain minerals of the rare low-atomic-number elements like lithium mentioned above. Rare earth pegmatites are those which contain groups of **high atomic number elements** (of the periodic table) like the rare earths (lanthanide series) and the actinide series. Such **pegmatites are known as rare earth pegmatites; they can be a source of these elements.**

Brazilian Pegmatites

Large areas of the Amazon River Basin are underlain by Precambrian rocks—granites as usual, being especially representative. Associated with these granites are pegmatites, especially in the Brazilian province of Minas Gerais. In a tropical, wet environment, granite (under chemical weathering) becomes deeply weathered and pegmatite minerals that contain lithium, beryllium, and boron become enhanced, while more common silicates, like the feldspars, become clay. The more exotic minerals, not being affected by this weathering, are released in perfect condition to yield gem minerals. Sometimes pegmatites in this deeply weathered granite are referred to as "gem pegmatites."

Pegmatite outcrop in northern Minnesota. This pegmatite contains small iron bearing, black tourmaline crystals where it forms part of an Archean granite. Pegmatite outcrops at high latitudes generally are unyielding of their crystals and its usually difficult to extract specimens from them. This is in contrast to outcrops in warmer climates where deep weathering of the pegmatites aids in the extraction of the gem crystals.

Close-up of Minnesota pegmatite crystals.

Minas Gerais Brazilian Pegmatite Crystals

A gallery of beautiful gem minerals from pegmatites of the province of Minas Gerais, Brazil.

A granite mountain, Minas Gerais, Brazil. The province of Minas Gerais in southeastern Brazil is predominantly a terrain composed of Precambrian granites. These granites have undergone deep chemical weathering in a wet, tropical climate. **Such weathering greatly aids in the extraction of pegmatite minerals, especially the hard silicates prized as gem minerals.**

Sugar Loaf, Rio de Janeiro, Brazil. This monadnock is composed of granite. Granite underlies a large portion of Brazil.

Entrance to a small pegmatite mine dug into red clay formed from chemically weathered granite in the wet and warm climate of Minas Gerais, Brazil. Granites under these conditions weather to red clay and pegmatites in them, when subjected to such deep weathering, have their feldspars converted to clay. Other silicates like tourmaline and beryl are not adversely affected by this weathering and sometimes are even enhanced by the process. Tropical weathering enables the crystals to be released fracture-free where they can be dug out in small mining operations like this. Many of the gem minerals of Brazil come from such sources. *Photo courtesy of Warren Wagner*

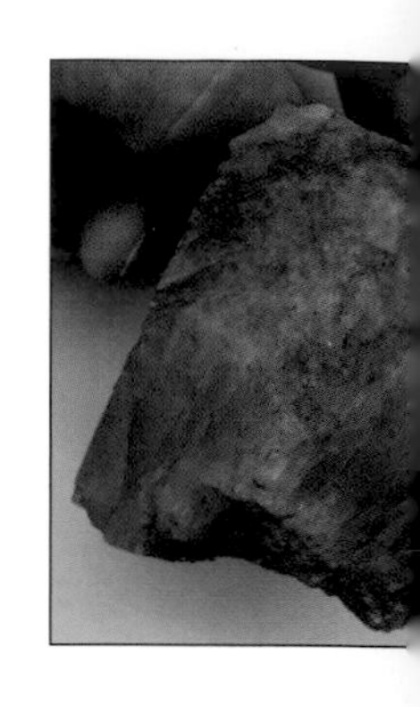

Beryl crystal from the mine shown in the previous photo. The red color on the otherwise blue-green beryl is red, iron rich clay, a stain coming from the chemical weathering processes characteristic of this tropical environment. Small gem pegmatite mines sometimes will have their crystals showing this red stain characteristic of deep weathering, a process which helps considerably to separate the crystals from the rock. Crystals from larger pegmatite bodies worked currently have weathered under similar circumstances, but usually lack such stains. The process of deep chemical weathering on gem minerals often imparts a gemmy aspect to the crystals, while at the same time allowing their easy extraction from the rock matrix. *Specimen courtesy of Warren Wagner*

Tourmaline, light green, Minas Gerais, Brazil. Light colored (green and blue) tourmaline is known as verdelite. Depending upon the absence of iron (which forms black tourmaline known as schorl), various other elements in tourmaline (like lithium and beryllium) produce a beautiful range of colors, sometimes with different colors showing in the same crystal. These and the following crystals came from large gem pegmatite bodies of Minas Gerais, Brazil. (Value range F).

Pink tourmaline, Minas Gerais, Brazil. Pink and red tourmaline are known as rubellite. Pegmatite minerals from relatively large pegmatites, which produced the following crystals from Minas Gerais, lack the reddish regolith stain. They came from deeper within a large pegmatite body where such stain has not migrated. (Value range G).

Schorl, black tourmaline. Tourmaline containing ferrous iron from the province of Minas Gerais. The pigmentation of iron overwhelms any color produced by rarer elements like lithium or beryllium. Schorl is a much more common form of tourmaline than either rubellite or elbaite. (Value range F).

Schorl. Large crystal of this iron bearing tourmaline from Minas Gerais.

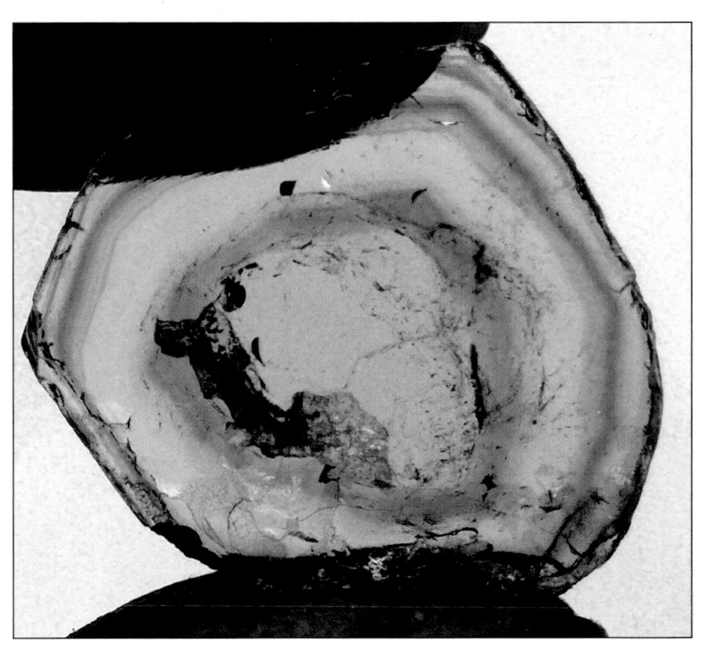

Elbaite slab. Iron free, unlike the previous specimen, the thin slice is known as watermelon tourmaline. It came from a fracture-free red and green tourmaline crystal. Tourmaline crystals usually have some fractures in them, a "flaw" that reduces their value in their use as gem stones. This slab is free of any such fractures, so it is more pricey. Province of Minas Gerais, Brazil. (Value range E).

Hematite. The presence of this culprit in a pegmatite will pretty much guarantee that tourmaline present nearby will be the iron bearing schorl. (Value range G).

Slender, light pink rubelite crystal (thumbnail size). An especially gemmy crystal from Brazil. (Value range F).

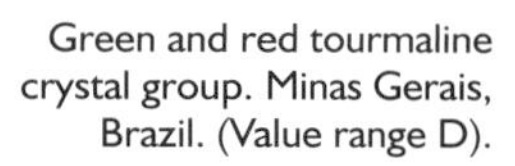

Green and red tourmaline crystal group. Minas Gerais, Brazil. (Value range D).

Dark green tourmaline with quartz. (Value range F).

Same group as shown in the previous photo, but with the crystal showing some pink tourmaline. Such red and green tourmalines are referred to as watermelon tourmaline.

Dark green tourmaline and quartz. Backside of previous specimen. (Value range F).

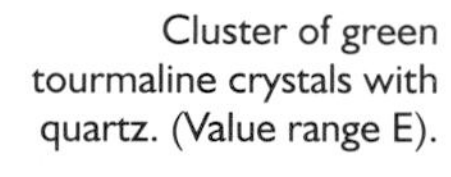

Cluster of green tourmaline crystals with quartz. (Value range E).

Pink tourmaline (rubellite) with quartz and feldspar. (Value range E).

Crystals of green tourmaline (verdelite, with some rubellite) to make a watermelon tourmaline group. (Value range E).

Elongate green tourmaline crystal set in quartz. (Value range F).

Fractured pink tourmaline. Extensive fracturing like this reduces the value of a specimen. Such fracturing probably accompanied the deep chemical weathering processes, which allows crystals to be removed from the pegmatitic mass without breaking them. (Value range G).

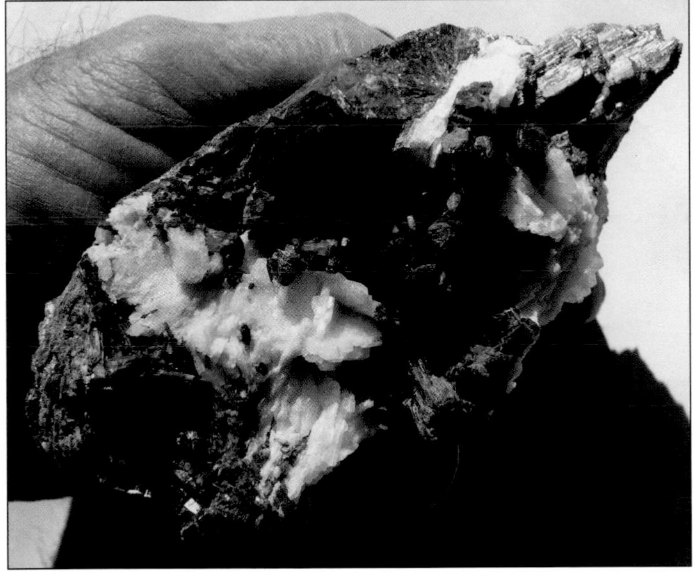

Cluster of dark green tourmaline crystals with quartz. (Value range E).

Fractured rubellite crystals. The more fractures, the less desirable. (Value range G).

Pink tourmaline (rubellite) with fractures. Province of Minas Gerais, Brazil. (Value range G).

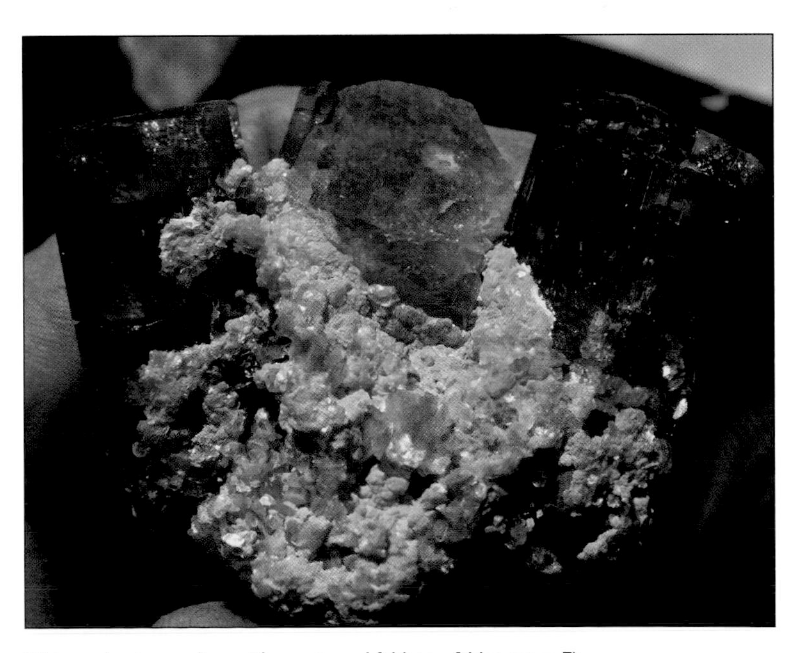

Watermelon tourmaline with quartz and feldspar. (Value range E).

Radiating cluster of grass-green tourmaline crystals in quartz and feldspar. (Value range F).

BRAZILIANITE
Linopolis,
Minas Gerais,
Brazil

Brazilianite, Linopolis, Minas Gerais. A hard phosphate mineral considered as a gem mineral. It comes from rare phosphate containing pegmatites.

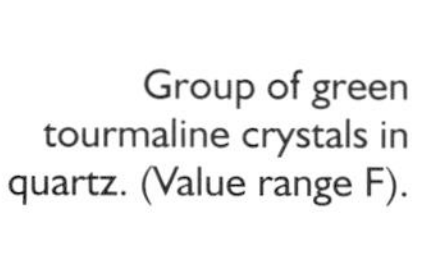

Group of green tourmaline crystals in quartz. (Value range F).

Light blue tourmaline. Less commonly seen than other colors, hence more desirable. (Value range E).

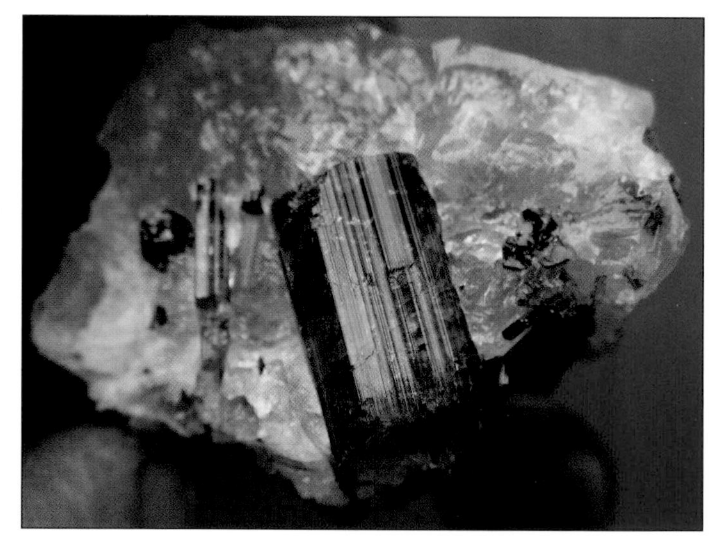

Large tourmaline crystal in quartz. Specimens like this, where the tourmaline crystal is snuggly placed in quartz, are often more desirable than is a stand-alone-crystal. However, this often depends upon the taste of the collector. (Value range F).

Scapolite. This is another silicate mineral from Brazilian pegmatites. It is rare in this environment.

Green tourmaline crystals in quartz. (Value range F).

Skorl. This plebeian tourmaline pales when compared with specimens like the one shown previously. (Value range G).

Kyanite. These attractive blue crystals are embedded in quartz. Kyanite is not common in pegmatites. It is usually associated with metamorphic rocks like schist or gneiss. It also is not usually this attractive. (Value range F).

Spectacular watermelon tourmaline (Value range D).

Brazilian blue kyanite.

Quartz crystal group. Excellent quartz crystals occur in the gem pegmatites of Minas Gerais. Well-formed quartz crystals are a phenomena associated with climates where deep weathering occurs. In pegmatites like those occurring on the Canadian Shield, inclusions in well formed, stand-alone quartz crystals like this are rare. As is the case with tourmaline and beryl, the deep chemical weathering of warm climates affects the pegmatite so that the crystals can be separated and extracted. (Value range F).

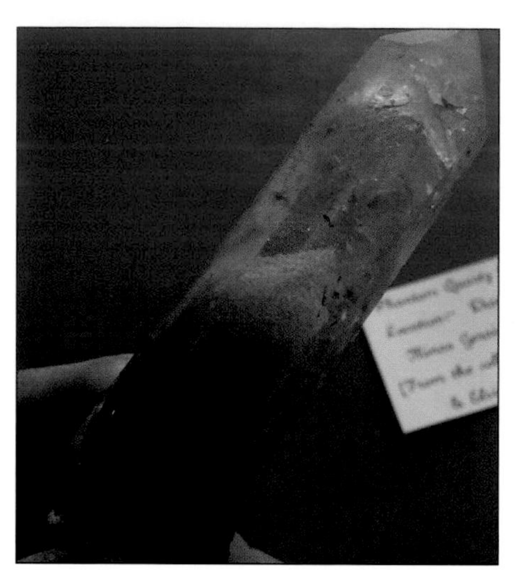

Quartz crystal with "sand." Small fragments of other minerals were incorporated into the quartz crystal when it grew. The growth of some of these quartz crystals may have taken place after the pegmatite was formed. Included material like this appears to be especially characteristic of pegmatites, which have undergone deep chemical weathering under a warm climate. (Value range F).

Inclusions in Pegmatite Related Quartz Crystals

Crystals of quartz associated with Brazilian gem pegmatites sometimes contain crystals of other minerals. Such a mineral embedded in another is known as an inclusion! Crystals of quartz containing inclusions may have formed after the pegmatite mass itself formed, possibly much later, after the granite mass containing the pegmatite was exposed at the earth's surface. Exposed after hundreds of millions of years of erosion to remove what originally covered the granite, some of these quartz crystals containing the inclusions may have formed during the weathering process itself, perhaps only a few millions of years ago. The weathering process, under certain conditions, appears to either form some quartz crystals or to allow them to grow larger and more "gemmy."

Quartz crystal group with black inclusions, Minas Gerais, Brazil. (Value range F).

Quartz crystal with ingrown tourmaline inclusions. (Value range F).

Quartz crystal on which small mica crystals and small quartz crystals have grown. (Value range G).

Quartz crystal containing inclusions of tourmaline. Quartz containing such an inclusion is known as tourmalated quartz. (Value range G).

Clear quartz with "sand." (Value range F).

Quartz crystals containing "crud"—flakes of quartz from within the pegmatite. Did these crystals form later, as in a geode? Where did the "crud" come from, within the pegmatite? (Value range F for group).

Short quartz crystals with "sand." (Value range F).

Smoky Quartz. (Value range F).

Quartz crystal with "sand or crud." (Value range G).

Rose Quartz

Pink, massive quartz can be associated with pegmatites like those of the Black Hills of South Dakota. In a few pegmatite occurrences in Brazil, actual rose quartz crystals are found. Otherwise, crystals of rose quartz are a rare occurrence.

Single quartz crystal with rutile needle inclusions. (Value range F).

Group of light-shaded rose quartz crystals. (Value range F).

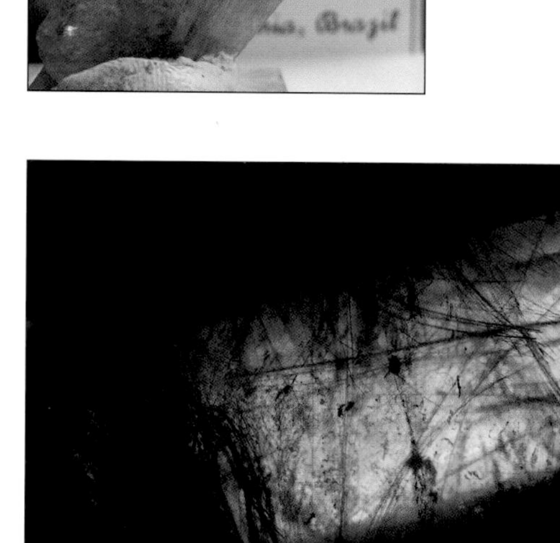

Polished quartz crystal with rutile crystals photographed by transmitted light. (Value range F).

Rose quartz crystals. Groups of rose quartz crystals, a mineral previously quite rare to be found as crystals, have come from Brazilian pegmatites starting around 1998. (Value range F).

Rutilated quartz. A nice group of rutile crystals lovingly encapsulated in quartz. (Value range F).

Group of large (40 mm in width) rutilated quartz crystals. Specimen originally in Ted and Elsie Boente's collection of St. Louis. (Value range E).

Top view of previous specimen.

Clear tourmalated quartz crystal. (Value range F).

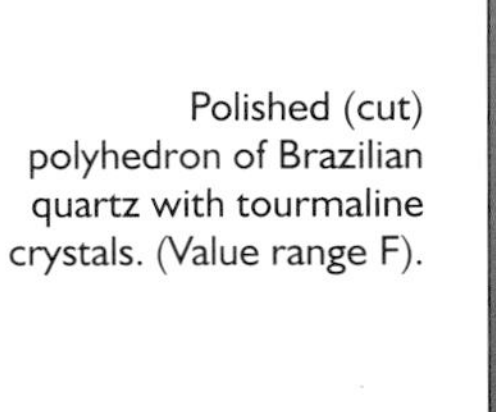

Polished (cut) polyhedron of Brazilian quartz with tourmaline crystals. (Value range F).

Polished "pebble" of clear quartz with rutile inclusions. A number of these highly polished specimens of rutilated quartz have entered the collector's market from China. The rutilated quartz, however, appears originally to have come from Brazil. (Value range F).

Faceted tourmalated quartz. (Value range F).

Polished clear quartz (cut) polyhedron with interesting rutile crystals. Many of these polished specimens from Brazil were recently placed on the mineral market that were polished in China. Some of these say "product of China," although the actual mineral specimen came from Brazil. (Value range F).

Rutile Crystals

Rutile is a form of titanium dioxide. It exists in pegmatites for the same reason the occurrences of lithium, beryllium, and boron minerals do. That is, because titanium atoms do not fit into the space-lattice of the common silicate minerals like feldspar so it is excluded when they form, when it then concentrates and forms its own minerals like rutile. Titanium occurs most frequently as the oxide. Rutile, however, can also be what is known as an accessory mineral in granite, where tiny crystals of it can occur in the granite itself. As a consequence of this, titanium is the tenth most abundant element in the earth's crust. These small crystals of rutile are released when the granite undergoes weathering and they then become a component of sand. The black sand grains in common sand often are rutile crystals, small crystals that can be separated from sand as a source of titanium metal and its compounds. Titanium is a widely used industrial element, especially in the aircraft industry and its purified oxide is extensively used as a white pigment.

Rutile crystals (yellow) on hematite (black), both attached to clear quartz (left). Ibitiara, Bahia, Brazil. (Value range F).

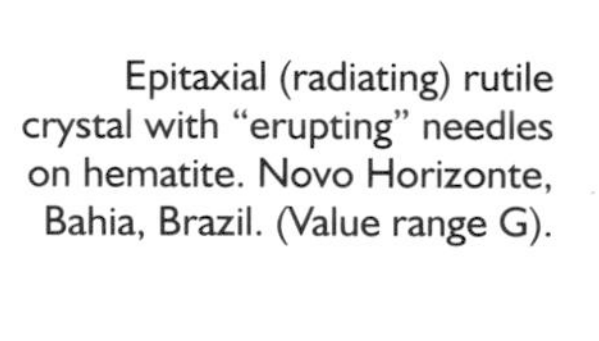

Epitaxial (radiating) rutile crystal with "erupting" needles on hematite. Novo Horizonte, Bahia, Brazil. (Value range G).

Rutile after anatase. (Value range G).

Cluster of rutile crystals on hematite, Minas Gerais, Brazil. (Value range G).

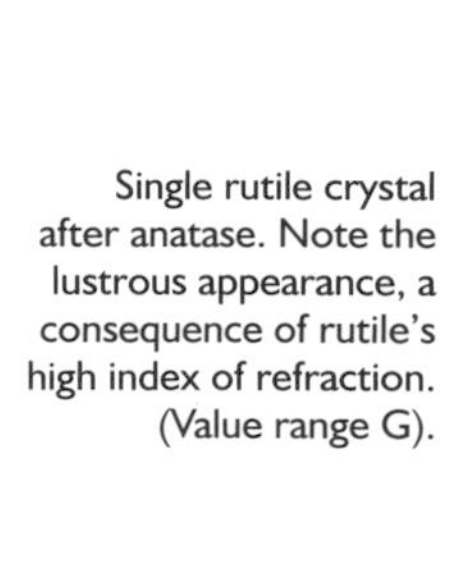

Single rutile crystal after anatase. Note the lustrous appearance, a consequence of rutile's high index of refraction. (Value range G).

Cluster of rutile crystals. (Value range G).

Beryl and Beryllium

Beryllium is one of the more peculiar elements of low atomic number (carbon is the first, with its property to form very complex, organic compounds, which are the basis for life). Not produced in the cores of stars, as were most other, common low atomic number elements, beryllium is rare. Beryl is the mineral in which beryllium is most concentrated and this is one of the more frequently found minerals in pegmatites.

Single euhedral rutile crystal. (Value range F).

Aquamarine. This gem mineral is a variety of beryl. Very nice aquamarine crystals occur in pegmatites on Mount Antero, Chaffee Co., Colorado (Value range F).

Single brookite crystal. (Value range G).

Another aquamarine crystal from Mount Antero, Colorado. (Value range F).

Beryl. This large beryl crystal came from the pegmatites of Crystal Peak, Colorado. It is a pegmatite body occurring in Pikes Peak granite. (Value range E).

Slender beryl crystals in feldspar and quartz matrix from central Wyoming.

Pikes Peak granite, Pikes Peak, Colorado. Large granite masses compose some of the highest peaks of the Rocky Mountains. This is a consequence of this massive, hard rock being resistant to erosion so that when the uplifting of the Rocky Mountains took place, the granite was pushed up without too much material being removed, as would have been the case with softer (and less massive) rock. Pegmatites within such a massive amount of granite can be the source of excellent minerals and crystals, as is the case with those of Crystal Peak.

Beryl crystal (variation aquamarine) in quartz. In contrast to beryl crystals from US pegmatites, these beryl crystals are extracted from the pegmatites with little difficulty, except for the effort of digging into weathered granite to find the pegmatite itself, which may be difficult. Such gemmy crystals may occur in a group in the pegmatite surrounded by weathered granite, which now is mostly red clay. Minas Gerais, Brazil. (Value range E).

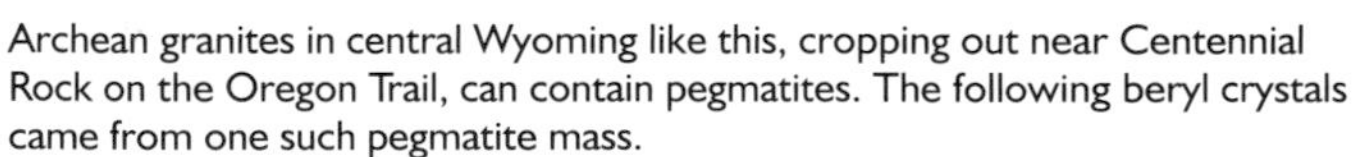

Archean granites in central Wyoming like this, cropping out near Centennial Rock on the Oregon Trail, can contain pegmatites. The following beryl crystals came from one such pegmatite mass.

Aquamarine crystal that has been totally extracted from any matrix. Minas Gerais, Brazil. (Value range E).

Beryl crystals collected by the author from north of Centennial Rock, Wyoming. (Value range E for the group).

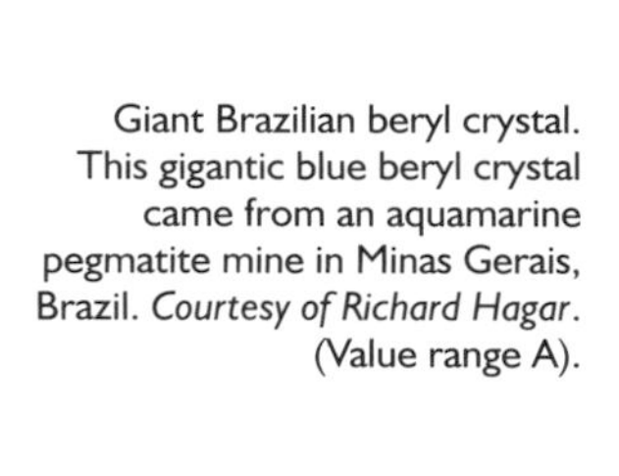

Giant Brazilian beryl crystal. This gigantic blue beryl crystal came from an aquamarine pegmatite mine in Minas Gerais, Brazil. *Courtesy of Richard Hagar*. (Value range A).

Aquamarine crystal with minor fractures. (Value range F).

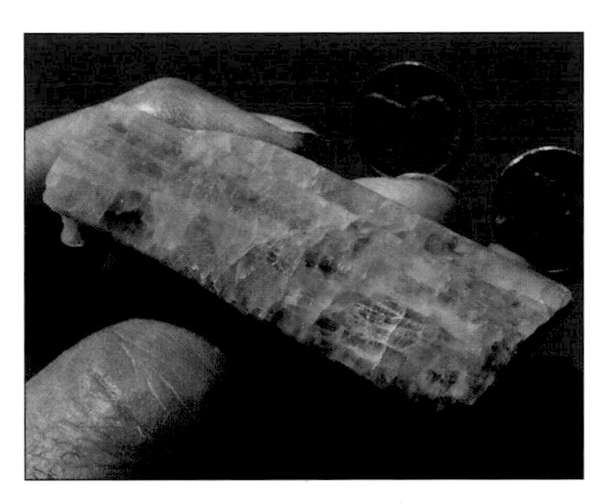

Aquamarine from India. Deep chemical weathering in India, like that of Brazil, can result in gemmy pegmatite minerals completely separated from the once enclosing rock. (Value range F).

Portion of aquamarine crystal. Conselheiro Pena, Minas Gerais, Brazil. (Value range F).

Aquamarine crystal in pegmatite, Afghanistan. Superb aquamarine crystals occur in pegmatite masses associated with younger, highly contorted sedimentary rocks. This complex geology represents part of the complex tectonic activity associated with the building of the Himalayan Mountains. Some of the nicest aquamarine crystals are associated with pegmatite masses that have become entangled with younger Paleozoic and Mesozoic rocks of Afghanistan. (Value range B).

Slender aquamarine crystal with black tourmaline in quartz from Afghanistan pegmatites. (Value range F).

Clear aquamarine. Most beryl and aquamarine is greenish or blue-green in color. This is clear and resembles quartz. Minas Gerais, Brazil. (Value range G).

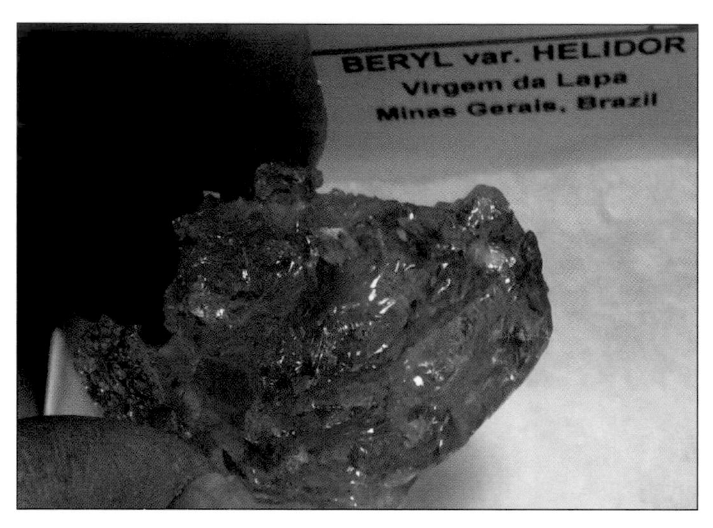

Beryl, variation Helidor. Beryl generally is greenish or light blue, as in the case of aquamarine. Helidor is a yellow form of beryl. Virgem da Lapa, Minas Gerais, Brazil. (Value range G).

Kunzite. A well formed, single crystal. Minas Gerais, Brazil. (Value range F).

Lithium Minerals

Lithium, like beryllium, is a low atomic number element, which is relatively rare. Lithium minerals, however, are more common than are those of beryllium. As with beryllium, most lithium minerals, like spodumene and lepidolite (a lithium mica), are associated with pegmatites. Lithium today is widely used in batteries for electronic equipment, like those used in cell phones.

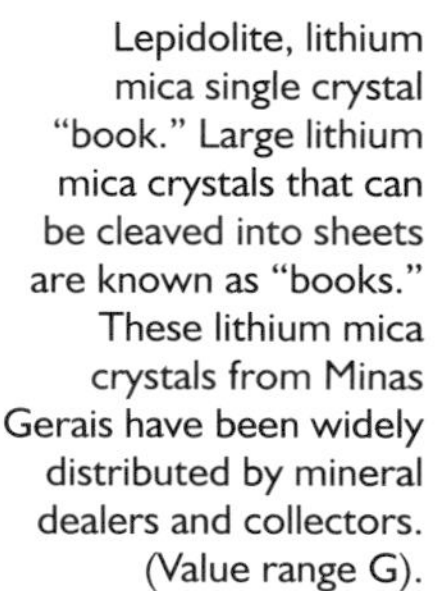

Lepidolite, lithium mica single crystal "book." Large lithium mica crystals that can be cleaved into sheets are known as "books." These lithium mica crystals from Minas Gerais have been widely distributed by mineral dealers and collectors. (Value range G).

Spodumene variation kunzite. Spodumene is a lithium silicate. Generally, it is opaque and resembles feldspar. A rare form of spodumene, known as kunzite, can be used as a gemstone. Gem quality spodumene comes from pegmatites of Minas Gerais, Brazil. (Value range G).

Lepidolite (aggregate of lithium mica crystals). Lepidolite is a frequently occurring mineral in pegmatites, especially gem pegmatites. Gem pegmatites influenced by deep weathering, such as occurs in the province of Minas Gerais, can have crystals undamaged after extraction from the pegmatite mass and thus are in a "perfect" condition. The three elements lithium, beryllium, and boron, composing minerals in pegmatites where iron is lacking, produce delicate colors in minerals, including the pink of lepidolite. (Value range H.)

Lepidolite crystals from a pegmatite in Siberia, Russia, made into a cabochon. Lepidolite is a mica found in ancient pegmatites worldwide, but most of it seen on the mineral market comes from Brazil. (Value range F).

Canoes offer an excellent way to geologize along waterways.

African Pegmatite Minerals

Large portions of sub-Saharan Africa are underlain by Precambrian rocks, much of them granite with pegmatites. Weathering in this region is similar to that of Brazil, where in a wet, tropical climate granite is turned into a red clay rich regolith. Pegmatite minerals, especially gemmy ones, can occur in this environment in a manner similar to that of Brazil.

Gravel bars can concentrate hard rocks like those coming from pegmatites. Minerals harder than quartz especially can survive weathering and concentrate in what sometimes are called gem gravels.

Red clay (terra rosa) derived from deep chemical weathering occurs widely over sub-Saharan Africa, as can be observed by these red alluvial sediments.

Topaz. This gemstone, considerably harder than quartz came from gem gravel in Africa. Note that it is abraded, but not severely. Softer minerals under these conditions would be totally abraded away. (Value range G).

Microcline and aegirine. Mount Malosa, Zomba District, Malawi. (Value range G).

Afghanistan Pegmatites

Precambrian pegmatites that have been inter-thrusted with younger rocks, in the complex geology of the Himalayan Mountains, yield beautiful gemmy crystals in Afghanistan.

Spessartine garnet in muscovite. Khugiani District, Nanngarhar Province, Afghanistan. The Afghanistan pegmatites yield exceptionally gemmy crystals, as well as unusual pegmatite occurrences, like this garnet. (Value range E).

Feldspar with black tourmaline (schorl). Erongo Mountain, Usakos and Omaruru districts, Erongo region, Namibia. (Value range F).

Cluster of verdelite tourmaline crystals sprinkled with quartz. Afghanistan tourmalines are some of the nicest ones to be found. Paprok, Kamdesh District, Nuristan Province, Afghanistan. (Value range B).

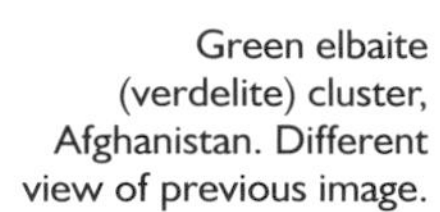

Green elbaite (verdelite) cluster, Afghanistan. Different view of previous image.

Elbaite crystal. Note slight pink at top and yellow at the bottom, a watermelon tourmaline. Nuristan Province, Afghanistan. (Value range B).

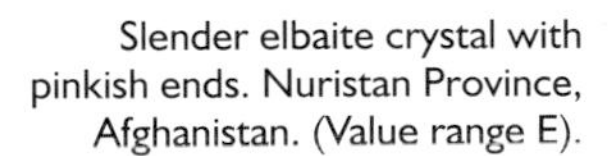

Slender elbaite crystal with pinkish ends. Nuristan Province, Afghanistan. (Value range E).

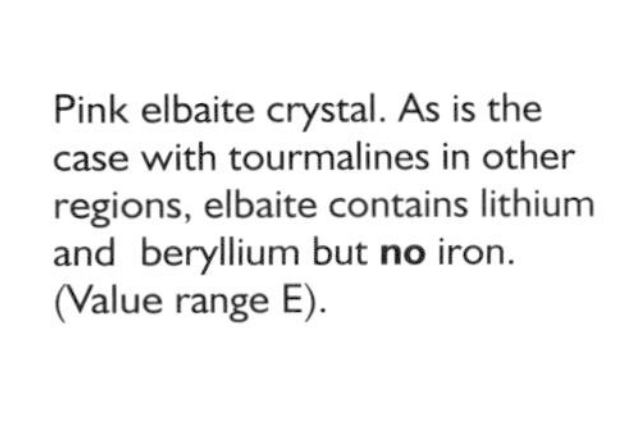

Pink elbaite crystal. As is the case with tourmalines in other regions, elbaite contains lithium and beryllium but **no** iron. (Value range E).

Some American Pegmatite Minerals

Pegmatites occur in considerable variety in the States—too much variety to do justice to them, however.

Schorl. Black tourmaline from pegmatites of North Carolina. (Value range G).

Epidote. A large crystal of a mineral rare in pegmatites. The southern Appalachian Mountains can locally have pegmatites in gneiss, which can contain some unusual minerals. Little Switzerland, North Carolina. (Value range F).

Pegmatite dike in Graniteville granite.

Graniteville, Missouri. Small pegmatites occur in some of the granites of the Missouri Ozarks, like here at Graniteville, Iron Co. Small pegmatite dikes occur in this pink granite which makes up the "pink elephants" at Missouri's Elephant Rock State Park.

Biotite mica from Missouri pegmatite.

Chapter Four

Some Early and Odd Mineral Occurrences

Ancient geologic terrains might be expected to be peculiar and different from those of later geologic time. Here are a few with their minerals, minerals formed during ancient orogenic (mountain building) events!

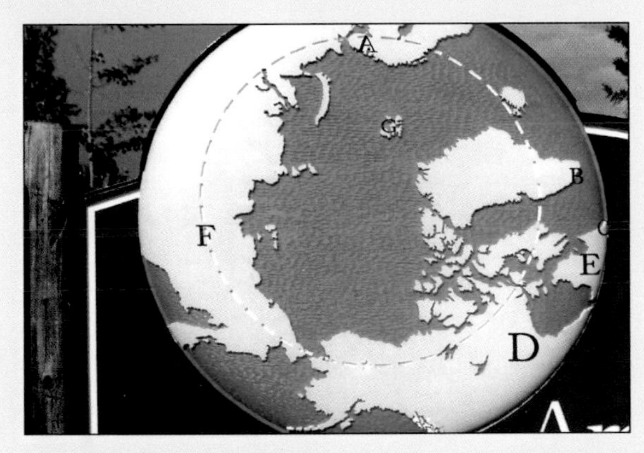

Map of the Northern Hemisphere as seen from the North Pole. Taken of a road sign on the Dalton Highway, northern Alaska at the Arctic Circle. A is the Kola Peninsula, B is southwest Greenland, C is coastal Labrador, D is west of Hudson's Bay in NWT Canada, E is Ungava (northern Quebec), F is Siberia, and G is Spitzbergen (Salbvard), an Island.

Large portions of this northern land can consist of muskeg or swamp. Both have lots of mosquitoes and black flies, but bedrock is often close to the surface or is exposed in glaciated outcrops in forested areas. Glaciated areas of hard Precambrian rocks have lots of depressions, which hold water. That is one of the reasons so many annoying insects, like mosquitoes, are so prevalent in northern shield areas.

The author in an area of ancient, early Precambrian (Archean) terrain south of Hudson's Bay, 1995.

Northern outcrops often are covered with lichens, but the rock itself is generally unweathered and fresh.

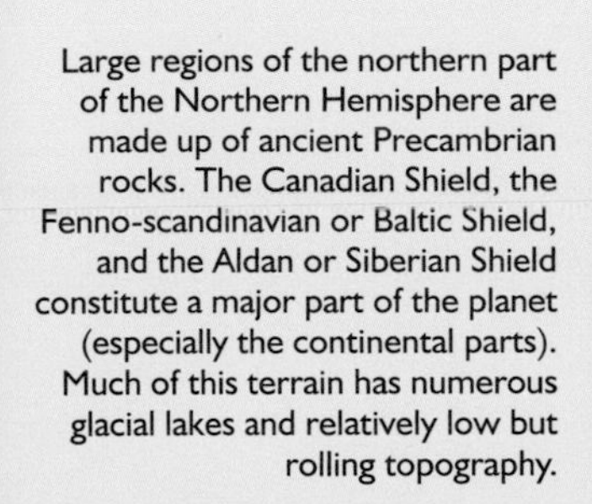

Large regions of the northern part of the Northern Hemisphere are made up of ancient Precambrian rocks. The Canadian Shield, the Fenno-scandinavian or Baltic Shield, and the Aldan or Siberian Shield constitute a major part of the planet (especially the continental parts). Much of this terrain has numerous glacial lakes and relatively low but rolling topography.

Bedrock in the north generally is fresh and unweathered; outcrops like this Archean gneiss were rounded by geologically recent glaciation. Outcrops at high latitudes commonly show such rounding from erosion by glaciers.

Lorenzenite, single crystal, Kola Peninsula, Russia.

Lorenzenite, same specimen—different view.

Kola Peninsula, Russia

An area of **early and mid-Archean** (ancient and somewhat anomalous) rocks comprises much of the Kola Peninsula, a bleak, rocky landscape in northwest Russia. This is an area that is part of the Fenno-Scandinavian Shield. The Kola Peninsula, which projects into the Arctic Ocean, is an Arctic landscape marked by A on the Arctic Circle road sign. Like many other high latitude regions of the Northern Hemisphere, it's a landscape composed to a major extent of Archean rocks. Part of the Kola Peninsula is underlain by nepheline-syenite, a relatively uncommon granite-like rock that, unlike granite, lacks quartz and is made up of peculiar silicate minerals. These minerals are deficient in silica and known as feldspathoids—**nepheline is a feldspathoid**. Associated with nepheline-syenite can be unusual minerals, some of which are both attractive and rare, and thus collectable.

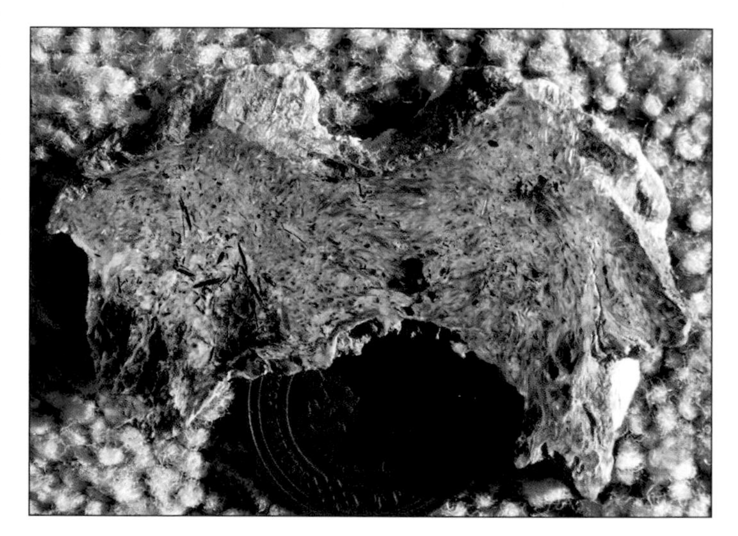

Yuksporite. A complex titanosilicate from the Kola Peninsula. Yuksporite is the pinkish-brown component. Titanium in this rare mineral substitutes for silicon. The mineral is named for Mt. Yuksporn where the specimen originated. (Value range F).

Lorenzenite from Archean rocks of the Kola Peninsula, in a matrix of nepheline. Lorenzenite is a sodium, titanium silicate. Lovozero Massif, Kola Peninsula, Russia.

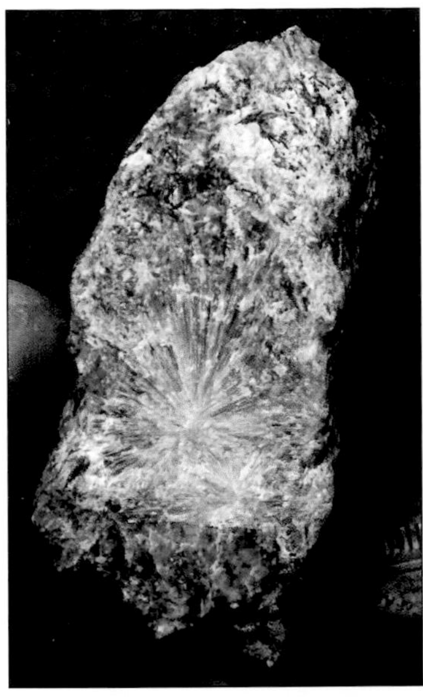

Titanite and Arfvedsonite. Mt. Yuksporn, Khibiny Massif, Kola Peninsula. (Arfvedsonite is the dark mineral; Titanite composes the radiating crystals).

Red corundum (ruby). A single crystal from the Kola Peninsula. Chit-Ostzov, Kola Peninsula. (Value range G).

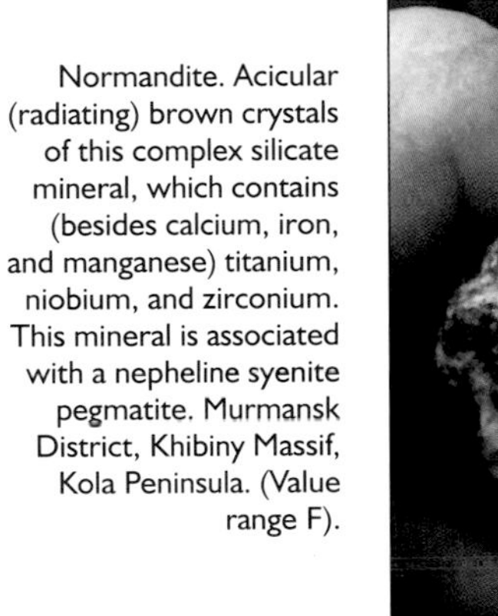

Normandite. Acicular (radiating) brown crystals of this complex silicate mineral, which contains (besides calcium, iron, and manganese) titanium, niobium, and zirconium. This mineral is associated with a nepheline syenite pegmatite. Murmansk District, Khibiny Massif, Kola Peninsula. (Value range F).

Eudialyte (reddish-purple mineral), sliced and polished slab. Eudialyte is a complex zirconium silicate mineral containing rare earths and chlorine. It is sometimes used as an unusual gem mineral. Khibiny, Kola Peninsula, Russia. (Value range F).

Astrophyllite in nepheline. Murmansk District, Khibiny Massif, Kola Peninsula. (Value range F).

Eudialyte. Fractured surface of small crystals of this unusual and attractive purplish-red mineral. Khibiny Massif, Kola Peninsula. (Value range G).

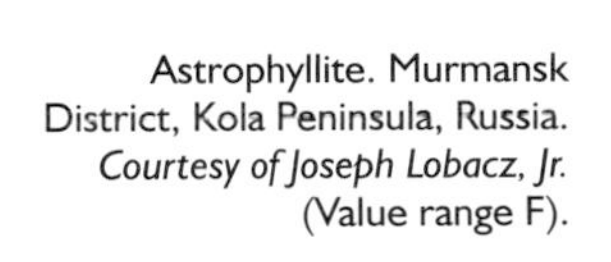

Astrophyllite. Murmansk District, Kola Peninsula, Russia. *Courtesy of Joseph Lobacz, Jr.* (Value range F).

Villiaumite. A rare sodium fluoride mineral. Koashua Mountain, Khibiny Massif, Kola Peninsula. (Value range F).

Amazonite (microcline). Parushaya Mt., Kejvy, Kola Peninsula. This mineral is associated with normal pegmatite, **not** with the quartz deficient types of many of the Kola Peninsula pegmatite occurrences. (Value range F).

Titanite (sphene). Brown crystals on mica schist. Khibiny Massif, Murmanskaya Oblast (District), Kola Peninsula, Russia. (Value range G).

Amazonite graphic granite. A slice of graphic granite from the Kola Peninsula (outer portion of a pegmatite) with amazonite as the feldspar. (Value range G).

Titanite (sphene) and hematite? Khibiny Massif, Kola Peninsula. (Value range G).

Carbonate-fluorapatite. Kovdor Mine, Kovdor Massif, Kola Peninsula, Russia. (An odd phosphate mineral). (Value range F).

Pyrrhotite. Nikolaevskiy Mine, Dalnegorsk, Kola Peninsula. (Value range G).

Almandine garnet. Kejvy, Kola Peninsula. A more commonly seen mineral associated with ancient (Archean) garnet schist of the Kola Peninsula. (Value range G).

Magnet Cove, Arkansas

Other areas of nepheline-syenite occur around the globe, which are also very ancient rocks like those in Jacupiranga, Brazil. A geologically younger region of nepheline-syenite, similar to that of the Kola Peninsula and Jacupiranga, occurs at Magnet Cove, Arkansas. Here, as in the Kola Peninsula, are found rare minerals associated with nepheline-syenite, some of which also are quite attractive and collectable.

Pectolite. A sodium, calcium silicate associated with high grade metamorphic rocks. Khibiny massif, Kola Peninsula. (Value range G).

Radiating crystals of titanite in calcite. This is from a carbonatite, a rare type of igneous calcite that forms a part of the Magnet Cove mineral complex. (Value range G).

Staurolite. Another relatively common mineral associated with high grade metamorphic terrains. It is a mineral that might be expected in this ancient terrain. (Value range G).

Apatite. Radiating or acicular crystals in carbonatite. (Value range G).

Charoite. A rare silicate mineral from Precambrian rocks of eastern Siberia. Charoite was discovered (or recognized) in 1978 where it formed from a syenite intruded marble in the Murunskii Massif (Aldan Shield). This occurrence is similar to some parts of the Kola Peninsula and to Magnet Cove, Arkansas. Specimen from the Chara River, Saha (Sakha) Republic, Yakutia, eastern Siberia, Russia. (Value range F).

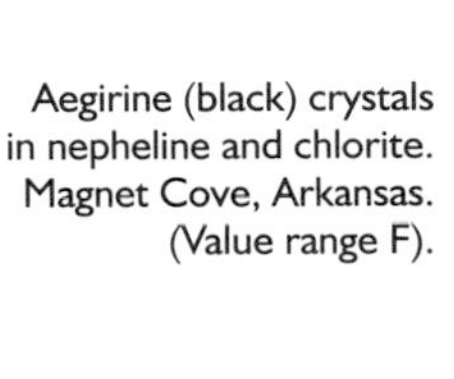

Aegirine (black) crystals in nepheline and chlorite. Magnet Cove, Arkansas. (Value range F).

Aegirine (or aegirite) in nepheline and chlorite. Magnet Cove Arkansas. (Value range F).

Close-up of previous shown rutile crystal.

Single Aegirine crystal, Magnet Cove, Arkansas. (Value range E).

Aegirine in nepheline. (Value range F).

Rutile. Aggregate of crystals. Magnet Cove, Spring Co., Arkansas. (Value range F).

Red ruby in nepheline and aegirine. The Magnet Cove mineral occurrence is considerably younger than is the Kola Peninsula, but in many other ways is similar to it. (Value range F).

Brookite crystal in carbonatite. Brookite is a rarer form of titanium dioxide than rutile. Both are found in the Magnet Cove area. (Value range F).

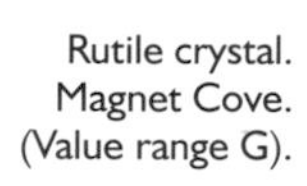

Rutile crystal. Magnet Cove. (Value range G).

Albite. Looking like barite, this feldspar is a common mineral in igneous rocks, Magnet Cove, Arkansas. (Value range G).

Aegirine in nepheline.

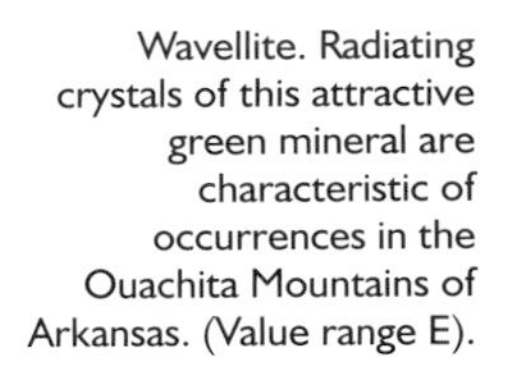

Wavellite. Radiating crystals of this attractive green mineral are characteristic of occurrences in the Ouachita Mountains of Arkansas. (Value range E).

Smoky quartz. The smoky quartz at Magnet Cove, Arkansas, probably was derived from exposure to natural radioactivity. Arkansas quartz found further west is artificially converted to smoky quartz by exposure to a source of intense radioactivity in a nuclear reactor. (The reactor at Missouri's S and T has often been used for this conversion).

Wavellite. This mineral occurs in the Magnet Cove area; however, nicer crystals are found farther to the west, especially near Mt. Ida, Arkansas. (Value range E).

Pyrite in carbonatite. Pyritohedrons embedded in what is igneous calcite, Magnet Cover, Arkansas.

Wavellite. Here wavellite has filled a fissure, and then the wavellite-filled fissure separated. The radiating crystals of this phosphate mineral are characteristic of the Arkansas occurrences. Mt. Ida, Arkansas. (Value range F).

Arkansas Phosphate Minerals

Unusual, slate-like, deep sea sedimentary rocks occur in the Ouachita Mountain region of central Arkansas, a region which includes the Magnet Cove area. Associated with these slaty rocks are a variety of phosphate minerals, some quite unusual, attractive, and rare.

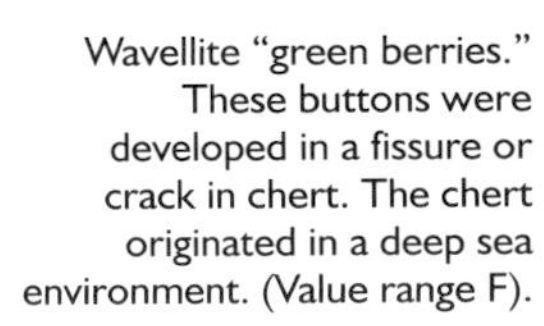

Wavellite "green berries." These buttons were developed in a fissure or crack in chert. The chert originated in a deep sea environment. (Value range F).

Kidwellite. A rare phosphate mineral named after Albert L. Kidwell (1919-2009). Kidwell had an active interest in, and was an authority on geologically anomalous areas of Arkansas and southern Missouri. He was especially attracted to the odd minerals found associated with geologically old (but not always geologically ancient) igneous intrusions of both the Ouachita and Ozark regions of those states. (Value range F).

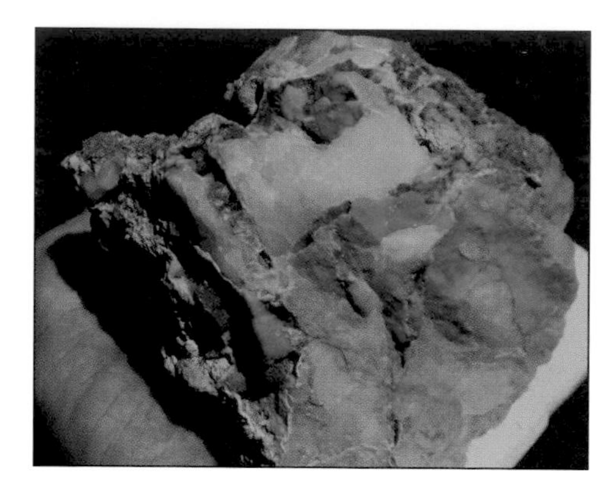
Varasite in quartz. This phosphate mineral is found in limited areas associated with wavellite. The best specimens come from west of Magnet Cove, but it occurs there also. (Value range F).

Varasite, Mt. Ida, Arkansas. (Value range F).

Keweenawan Basalts

A series of late Precambrian rocks containing puzzling mineral deposits occurs in the area surrounding Lake Superior in Canada and the United States, especially in Michigan's Keweenaw peninsula. Keweenawan rocks are primarily basalts and gabbro in the lower layers and reddish sediments in the upper portions of the sequence. These rocks are associated with what is known as the Lake Superior Rift, a geologically distinct portion of North America. A billion years ago, the Lake Superior Rift was placed in tension, resulting in the formation of a **rift zone** from which **large amounts of basaltic lava were extruded**. This rift zone resulted in a depression that, much later during the Pleistocene (Ice Age), was excavated by the movement of continental glaciers. Glaciers dug out the depression that then became the location of Lake Superior.

Close-up of red siltstone beds overlain by a Keweenawan basalt flow.

Outcrop of Keweenawan basalt sandwiched between red siltstone beds. North shore of Lake Superior, Ontario.

Prominent basalt flow overlaying Keweenawan red beds.

Gorge of the St. Croix River composed of Keweenawan basalt. Taylor's Falls, Minnesota. The St. Croix (a northern tributary of the Mississippi), upon encountering the hard Keweenawan basalts, cut this gorge with difficulty. The river is still cutting into this basalt—it's what makes Taylor's Falls, the falls on the St. Croix River, after which the town is named.

Lake Superior agates, Grand River, northwest Missouri. These agates were transported to northern Missouri by ice age glaciers from the region of Lake Superior. (Value range E for group).

St. Croix River gorge below Taylor's Falls on the St. Croix River.

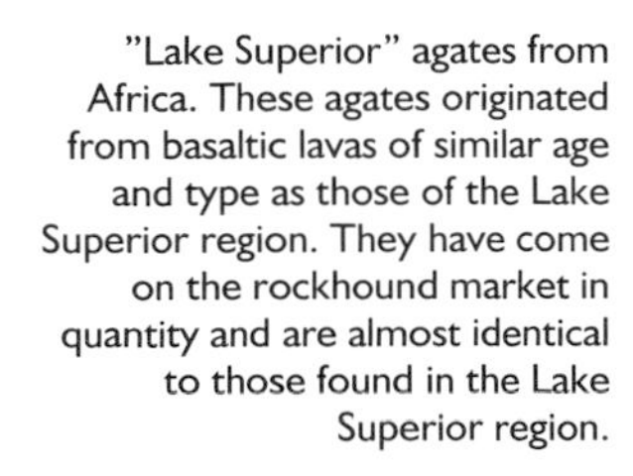

"Lake Superior" agates from Africa. These agates originated from basaltic lavas of similar age and type as those of the Lake Superior region. They have come on the rockhound market in quantity and are almost identical to those found in the Lake Superior region.

Keweenaw Peninsula Native Copper

A number of interesting minerals (and gemstones) occur associated with the mafic rocks of the Lake Superior rift zone, which include large quantities of native copper. The copper region in northern Michigan is locally known as Michigan's "copper country." This occurrence, one of the planet's largest (known) deposits of native copper, is geologically anomalous.

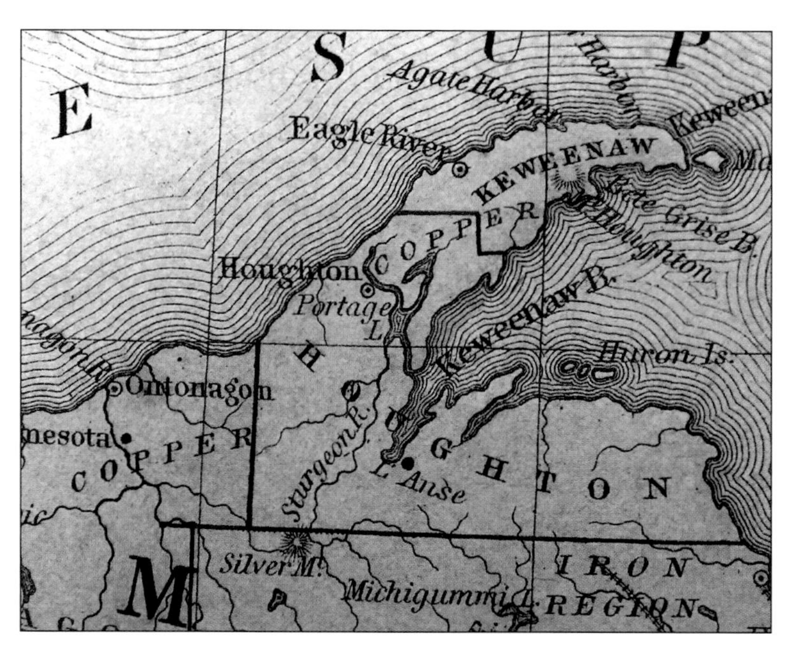

Map of the Keweenawan Peninsula, northern Michigan.

Michigan float copper. Native copper nuggets are found occasionally in glacial drift south and southwest of the Keweenawan Peninsula from which they originated. (Value range F).

Native copper mass collected from glacial till. The copper cobble has been sliced, polished, and coated. (Value range F).

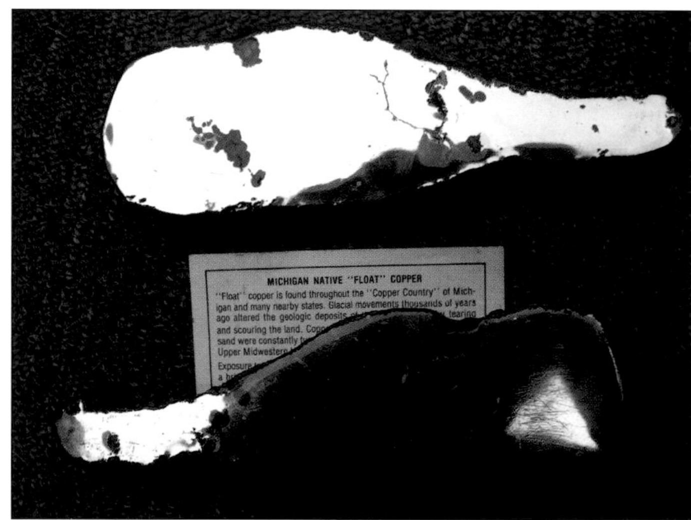

Two slices of native copper taken from a cobble of copper found as a glacial erratic in Indiana.

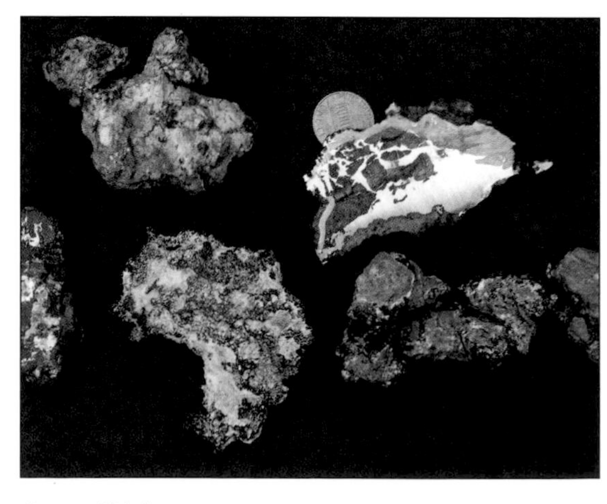

Group of Michigan native copper specimens. (Value range E for group).

Native copper in greenstone. Different age and types of rocks are mineralized with native copper in the Keweenawan Peninsula of northern Michigan. This is Archean greenstone into which the copper has been introduced. (Value range F).

Native copper nuggets with native silver (half breed). *Glenn Williams collection.*

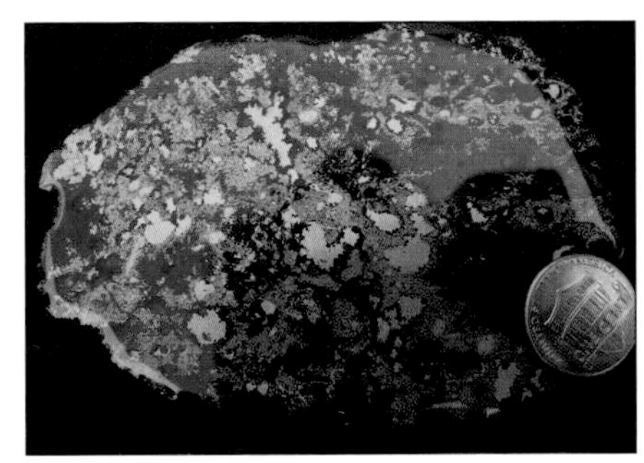

Slab of native copper. Polishing native copper is only a temporary arrangement, as it will corrode and tarnish in a few months. Copper with a clear plastic coating fares better, corrosion is stopped or arrested by this coating and the slab will retain a clean copper appearance. (Value range F).

Close-up of copper drift cobble containing native silver as well as copper. Courtesy of Glenn Williams.

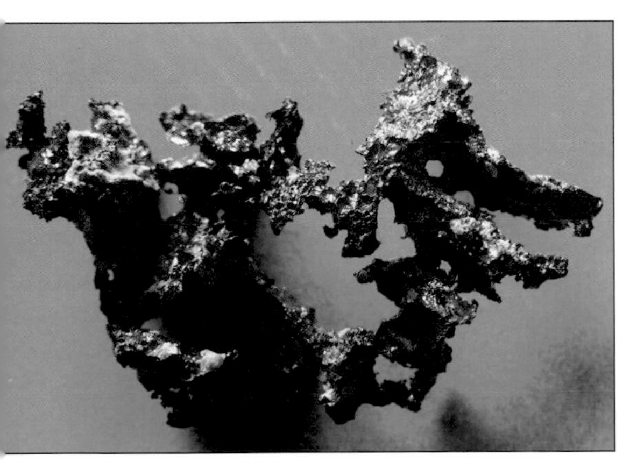

Copper mass from mines in the Keweenawan Peninsula with all matrix of volcanic rock removed. (Value range G).

Sheet copper. Native copper not only is associated with igneous rock in Michigan's Keweenawan Peninsula, but is also found associated with sedimentary rocks. Here is a slab of black, carbonaceous shale (Nonesuch Shale) with a thin sheet of native copper occurring along a bedding plane. (Value range E).

Another plastic coated slab of native copper in greenstone? (Value range F).

Stillwater Complex, Montana

A well exposed sequence of unusual Archean rocks occurs in part of the Absaroka Range of south-central Montana and is known as the Stillwater Complex. The Stillwater Complex consists of a series of mafic rocks making up what is known as a layered intrusive. In a layered intrusive, different types of crystals formed at different times and these are separated into layers—like the layers of sedimentary rocks. In the Stillwater Complex, pyroxenes can form one layer, amphibole yet another, creating a complex of alternating layers of different minerals. One of the economic aspects of the complex is the presence of mineable layers of the mineral chromite, a mineral sometimes characteristic of Archean terrains. Another mineable layer contains a concentration the rare element platinum, as well as other platinum group metals. The Stillwater Complex is a major source of platinum metal in North America.

Bronzite. Bronzite is a pyroxene often associated with stony (chondritic) meteorites. It is found in the Stillwater Complex, as these ancient mafic intrusions may have had their silicates derived directly from molten chondritic material. Bronzite and related minerals like hypersthene and enstatite (both also often major components of stony meteorites) are much rarer in younger mafic igneous rocks than in older, especially as large crystals. Bronzite contain some iron substituting for magnesium—the source of its sub-metallic sheen are schiller. (Value range G).

Chromite mine in the Absoraka Range, Montana, at 10,000+ feet.

Chromite (chromium oxide) from the Stillwater Complex.

Clinochlore (Seraphinite). A very hard form of the mineral chlorite, which sometimes is used as a gemstone. Archean rocks in Russia yield a number of interesting minerals like this, which are attractive and are used in jewelry. Eastern Siberia, deposit Korshunkovsky. (Value range F).

Some Minerals Associated with Ancient Orogenic (Mountain Building) Events

Ancient geologic terrains can yield a variety of (sometimes) unusual minerals; here are a few of them.

Pyrrhotite crystal with quartz. Pyrrhotite is a mineral closely related to trolite, an iron sulfide found associated with metallic meteorites. Pyrrhotite, unlike pyrite, chemically has the proper amount of sulfur (stoichiometrically correct), pyrite has an excess of sulfur. Pyrrhotite is also the iron sulfide associated with very ancient rocks, especially Archean greywackes.

Purpurite. A rare manganese phosphate associated with Archean greywacke. Larimar Co., Colorado. (Value range F for group).

Cummingtonite in schist. A rare brown silicate mineral found associated with high grade and ancient metamorphic rocks of the Black Hills of South Dakota. (Value range G).

Pumpellite. A rare mineral named after Raphael Pumpelly (1837-1923). Pumpelly worked in the late nineteenth century in Missouri, then focused his geological contributions on the Canadian Shield, where this silicate mineral occurs in Minnesota. Pumpellite is generally associated with early Precambrian rocks. (Value range F).

Archean Sulfide Minerals

The sulfide minerals shown here (chalcopyrite and bornite) were introduced during an orogenic (mountain building) event during the early Precambrian. When metamorphism of greywacke took place, mineralization took place deep within the crust.

Traveling along an elongate lineation (part of Nonacho Lake) by helicopter.

Metamorphosed Archean greywacke (meta-greywacke). What was originally "dirty sandstone" can be a frequently occurring rock of the ancient earth. Like other ancient rocks, it can be shot full of quartz veins, some of which can carry economically valuable minerals, especially the sulfides of silver and copper. It can also be a host for native gold.

Chalcopyrite with bornite and quartz in metagraywacke. Nonacho Lake, NWT Canada. Chalcopyrite is a copper-iron sulfide. In Archean terrains like this it can also be gold bearing.

Mineralized lineation. This elongate lake occupies a zone of weaker rocks that were either ground down by glaciers or were less resistant to weathering. This lineation is on the Canadian Shield. Such liniments can be the site of mineralization. Mineral deposits, like sulfides, can weather or make rocks weaker to weathering, and such soft rock was removed by continental glaciers during the last ice age. This lineation was prospected for both uranium and sulfide minerals. They both occur in the area but, as far as is known, only in non-economic quantities.

Chalcopyrite chunk. Nonacho Lake, Canada.

Bibliography

Wilson, Marc & George Robinson, 2008. "Copper Country." In *American Mineral Treasures*. East Hampton, CT: Lithographie.

Chapter Five

Precambrian Sediments and the Early Biosphere

Introduction

Life is an ancient phenomenon on the earth. Stromatolites and associated microfossils prove this, even taking into account scientific skepticism as to the oldest of these being of non-biogenic origin! Stromatolites, the subject of the following chapter, are structures built by primitive, prokaryotic life forms (cyanobacteria), usually in shallow water. The oldest stromatolites go back to 3.5 Gy (giga or billion years). Sediments (now metamorphosed) **suggestive of the presence of life** go back to almost **four billion years**.

Graphitic or graphite bearing schist. High pressure accompanying deep burial in the earth's crust changed (metamorphosed) a carbonaceous (carbon-rich) shale into this crystalline mass. Many Archean rocks are graphite bearing like this, the original carbon coming either from primitive life or from a pre-life, organic "soup" called a coacervate.

Biogenic Sediments I, Graphitic Slates, Phyllites, and Schist

One of the aspects of many Archean sedimentary rocks (now usually severely metamorphosed) which formed during the early earth is their dark appearance. Sometimes this is because they contain an abundance of carbon—carbon usually occurring as graphite. Graphite schist (or gneiss) can be one of these ancient rocks. The carbon they contain was possibly associated with life when they were deposited.

It has been stated by some workers regarding the early fossil and rock record (Elso Barghoorn, personal communication) that some 75% of the earth's carbon budget is locked up in the form of graphitic gneiss and schist in Archean metasediments. This can be appreciated as Archean metasediments in many regions are black or dark colored, usually due to the presence of carbon. The reason this is interesting and important is because the presence of dark material (either as graphite or as finely disseminated iron sulfide) usually implies the presence of reducing or anoxic conditions present where the sediment was deposited (usually in some ancient ocean).

A mixed bag of Archean graphitic slate which has been intruded by granite near Thermopolis, Hot Springs County, Wyoming. Such rocks as these have been preserved over vast spans of geologic time by being buried deep in the earths crust, which thus protected them from weathering and erosion for over three billion years. In the process of deep burial, they were intruded (injected) by granite, which formed from magma ultimately coming from the earth's mantle. The dark colored layered rock contains carbon in the form of graphite, the presence of which may be from a biogenic source. This form of graphite may also be from some type of organic "soup" (coacervate) that existed in the sea before life appeared. These outcrops form part of the Wyoming Province of the North American Precambrian craton—one of the oldest portions of North America.

Reducing or anoxic conditions indicate the absence of free oxygen—a condition which appears to be the norm with both Hadean and Archean Earth. An anoxic environment is also necessary as a prerequisite for organic chemical evolution—the mechanism favored by scientists who extend Darwinian natural selection backward (in the form of chemical evolution) through geologic time into the pre-biotic earth.

The prevalence of graphite in these early rocks has also, however, been explained as its being a residue of inorganic carbon compounds generated from what is known as a coecervate. The concept of a coeacervate is a hypothetical position, one which presumes that large amounts of organic compounds were produced during the early earth from chemical reactions similar to those generated in the Miller Experiment. These would consist of hydrocarbons, aldehydes, amino acids, and many other organic molecules. Indeed, the reddish and brownish material underneath the surface of Jupiter and similar material detected on Saturn's satellite Titan appears to have formed in this manner (also the organic material found in carbonaceous chondritic meteorites probably formed in a similar way in a nebula). If it indeed formed from a coacervate, this carbon residue in the oldest of rocks would be from **pre-biotic conditions**, that is before there was any life on Earth. This would apply only to the earliest of terrestrial records, as life itself appears to be a very ancient phenomenon on the planet. Stromatolites as old as 3.5 billion years are known from Archean rocks (see chapter six).

The matter of these carbon-rich slates inevitably gets involved with the issue of the origin of life itself, a subject wrought with controversy and beyond the scope or focus of this work. This extremely interesting subject can be summarized by noting the following hypotheses as to life's ultimate origin or introduction—a phenomenon that did take place within the time frame covered by this work:

Organic Soup: Life originated from chemical evolution involving carbon containing compounds in an organic-compound-rich "soup" (a coacervate). Such an "organic soup" originated in the earth's oceans in a reducing environment from reactions similar to those of the Miller Experiment.

Panspermia: Life originated from some place in space other than the earth and "seeded" itself through some sort of minute "life spores" that may have ranged throughout the galaxy. This (somewhat outlandish appearing hypothesis) proposes that small packets of DNA were (or are) scattered throughout the Milky Way Galaxy, taking hold if they landed in a suitable place (like the earth) where they then reproduced and evolved.

Directed Panspermia: This hypothesis offers that life was introduced to the earth from intelligent space travelers early in its history. Here they either intentionally or unintentionally left behind "life spores" in the form of DNA or prokaryotic cells. It might be mentioned that both forms of panspermia were originally proposed by molecular biologists to explain the early appearance and complexity of life at the cellular level. Chemical evolution on the earth did not have sufficient time to produce the complex molecular precursors of life, such as DNA.

God Hypothesis: That life is a product of some sort of divine or supernatural action.

Black slate, black from both iron sulfide and graphite, Black Hills, South Dakota. These black slates are a typical Archean rock.

Open-pit portion of the Homestake gold mine, Lead, South Dakota. Archean rocks can be auriferous (gold bearing) as are these black, graphite rich slates mined extensively for their gold content until recently.

Vertical Archean strata exposed in western Ontario along the Trans-Canadian Highway. The vertical beds are composed of black slates and greywacke. Archean strata are usually vertical like shown here, a consequence of their original deposition on the sea floor and later association with upward movement of masses of granite from below the earth's crust.

Black, iron bearing dolomite and greywacke exposed in an open pit iron mine near Wawa, Ontario, along Lake Superior. Archean sediments often are high in iron compounds. This is a type of iron formation, although most of the iron is present as iron carbonate and not hematite, which is the usual form of iron in algoman-type iron formation. This is the type of iron formation characteristic of the Archean.

Biogenic? Sediments-Banded Iron Formation (BIF)

What has been known for years as a peculiar type of rock limited to the Precambrian is known as banded iron formation or BIF. It is usually made up of layers of hematite alternating with layers of jasper (a mixture of hematite and quartz) and quartz (usually as a form of chert). Often the layers in BIF are distinct. Such distinct bedding usually is not seen in younger sedimentary rocks, especially in sedimentary rocks deposited under marine conditions (as were most iron formations). Iron formation, being a rock type unique to the Precambrian, reflects a process that is also unique to that time. Iron formation is generally attributed to the evolution of the earth's atmosphere from an anoxic one to one containing free or elemental oxygen.

What constitutes iron formation is (sometimes) somewhat debatable. Precambrian sedimentary rocks often contain higher amounts of iron minerals than are found in similar rock types (like dolomite) of younger geologic age. This iron was chemically precipitated from bodies of water (usually marine), which apparently contained amounts of dissolved ferrous iron in solution in amounts not present in the Phanerozoic. A concentration of ferrous iron in sea water would be expected if atmospheric oxygen was either absent or was at a considerably lower level than exists today. Absence of oxygen in the atmosphere would allow iron to exist in large quantities in the ferrous oxidization state, which unlike **ferric iron** is highly soluble in water. When oxygen encounters ferrous iron, it chemically joins up with it, producing ferric iron, that insoluble form of iron found abundantly in iron formation.

Close-up of open pit mine exposing Archean iron formation at Wawa Ontario.

Metamorphosed iron formation exposed near Flin Flon in west central Manitoba. The beds of iron formation here are vertical as is typical of Archean terrains. This occurrence is similar to the metamorphosed iron formation in the 3.9 billion year old Isua Series of southwestern Greenland, some of the oldest rock on the planet.

Algoma-type iron formation known as the Soudan Iron Formation. It is part of the 3.0 billion year old Ely Greenstone Belt of northern Minnesota.

Chunk of Algoma-type iron formation set in a masonry wall. Soudan Iron Mine, northern Minnesota.

The Soudan Iron Mine, a former underground iron mine that worked the Soudan Iron Formation near Ely, Minnesota. The mine is now used as an underground neutrino detector.

Two and a half billion year old conglomerate (a rock type made up of cemented pebbles). One of the pebbles in the conglomerate is a piece of Algoma-type iron formation. Such a pebble was derived by weathering from an outcrop that existed 2.5 billion years ago. It was then carried by a stream. This pebble, along with others, was then deposited to form a gravel deposit, which eventually became the conglomerate. The age of the pebble has to be millions of years older than the conglomerate in which it is embedded (and this conglomerate is 2.5 billion years old). From outcrops at the eastern end of Lake Superior, Wawa, Ontario.

Plaque mounted on a boulder of Algoma-type iron formation of the Archean Era from a former iron mine in western Ontario. Algoma-type iron formation of the Archean Era around Lake Superior have been extensively worked as a source of iron ore for over 100 years. It occurs in the area of Lake Superior that is some of the older terrain of the Canadian Shield and known as the Superior Province.

Conglomerate of early Proterozoic age (approx. 2.4 billion years old) containing small jasper pebbles or clasts derived from a much older outcrop of iron formation.

Pebbles of Algoma-type iron formation (probably derived from late Archean Greenstone Belts of Minnesota). These pebbles were collected from a gravel bar of a river in northeastern Missouri. They can be common locally depending upon where the glacial drift of the area was derived. Pleistocene glaciers carried them southward in large quantities into Iowa, northern Missouri, Illinois, and Indiana. Rockhounds know such iron formation pebbles as jasper and sometimes work them into interesting jewelry. (Value range G).

Iron formation egg and heart exhibiting what are believed to be stromatolitic laminae. They rest on a slab of black chert that came from near Schrieber, Ontario, from Archean strata. Identical patterns suggest a common origin for both of them, yet they came from half a world apart. They also appear to be about the same geologic age, some three billion years old. (Value range G for each).

Iron formation slab from an isolated outcrop in central Wisconsin. The pattern on this sliced and polished slab suggests its an Archean iron formation, algoma–type. (Value range G).

Close-up of the same egg and heart as in the previous photo. A large number of these iron formation eggs are made from late Archean Iron Formation from Western Australia and polished in China—they have come through the Tucson, Arizona, show. (Value range G for both).

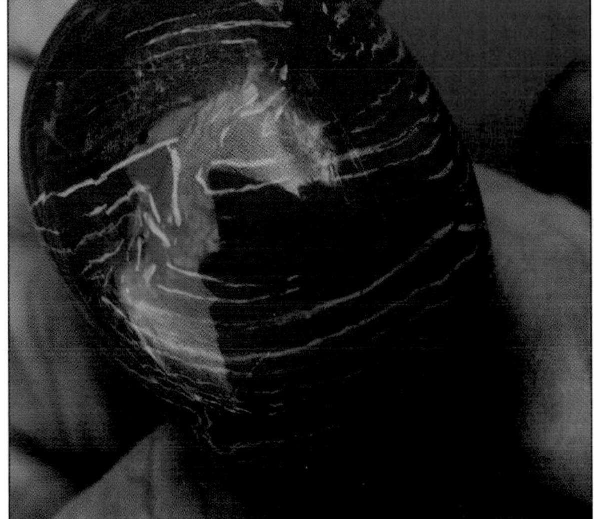

Iron Formation egg from Western Australia with distinct laminae. The laminae are suggestive of laminar stromatolites. Similar black iron formation from the late Archean of western Ontario has yielded microfossils of presumed monerans. (Value range G).

Group of spheres and eggs made from Archean iron formation (jasper) and polished in China. Specimens that came through the Tuscan show often will say "Made in China" but this is actually where the objects were polished from jasper imported from Western Australia.

Slab of BIF from Western Australia with stromatolite-like laminae.

Polished slab of BIF (banded iron formation) from Western Australia containing stromatolitic-like laminae. (Value range F).

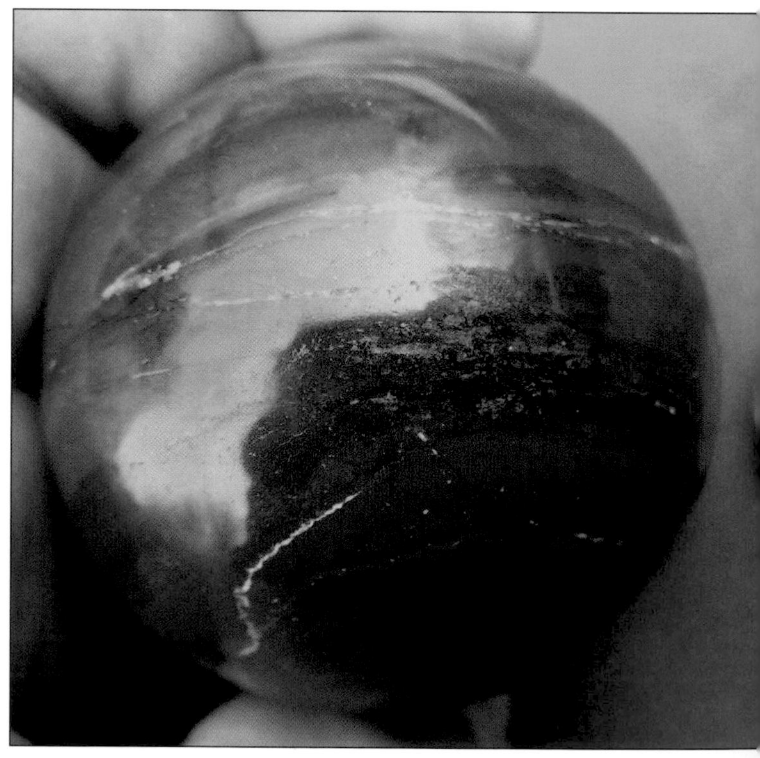

Sphere made from typical Algoma-type BIF. (Value range G).

Large (heavy) chunk of BIF from Western Australia. Known as tiger iron, this iron formation contains bands of tiger's eye, a form of quartz that has replaced asbestos and found associated with some Algoma-type iron formations. This material has been widely distributed among rockhounds who polish it into interesting jewelry.

Polished heart made of Western Australia (Hamersley iron district) algoma type iron formation. A large portion of Western Australia is underlain by early Precambrian age rocks. (Value range G).

Polished slice of tiger iron BIF. The yellow bands are tiger's eye, the red is jasper, and the blueish bands are hematite. From the Hamersley iron district of Western Australia. (Value range F).

Slab of BIF (tiger iron) containing a thick layer of tiger's eye from Western Australia (Value range F).

The back side of the previously shown slab with tiger's eye.

BIF from central Wisconsin. South of Lake Superior, a region of Archean rocks occurs in central Wisconsin. This BIF has profuse red and yellow jasper layers. It is a popular type of rock often worked into colorful jewelry by rockhounds. (Value range F).

Colorful slab of "tiger iron," Algoma type iron formation from the Hamersley Range, Western Australia.

BIF with both red and yellow tiger's eye. From Western Australia's Hamersley Iron range.

Binghamite (from Crosby-Ironton Minnesota)! This BIF is predominantly made up of quartz. The hematite layers have a luster like tiger's eye. Binghamite is a rockhound term. Rockhounds often give specific names to attractive and distinctive rocks, especially if these can be made into polished jewelry. Binghamite is a distinctive type of BIF. Some geologists frown on this practice as this nomenclatural procedure is not being done through the "official" protocol of geology. The author, himself a geologist, considers this appropriate as it serves to identify a rock that otherwise would not be recognized. This habit of rockhounds emphasizes and identifies a distinctive rock! A similar situation exists with mozarkite, a colorful form of quartz (flint), which is also distinctive and found over the Missouri Ozarks. Mozarkite was involved in a situation that caused some geologists heartburn when mozarkite was established by the Missouri legislature as the "official" state rock. (Value range F).

Mozarkite: The *official* state rock of Missouri. Although not from Precambrian rocks (but it comes close), mozarkite, like Binghamite (both named by rockhounds), is attractive and used in jewelry making. (Value range F for group).

Peculiar metamorphosed iron formation? From central Wisconsin, this is an iron-rich rock popular with rockhounds. It comes from Archean terrain. It represents one of many odd (and attractive) iron-rich rocks found in Archean terrains. (Value range F).

Pilbara Picasso Jasper. Looking similar to the jasper from Wisconsin shown in previous photos, this name is more of a rock dealer's creation than a rockhound identification. It comes from Western Australia in the Pilbara region, an extremely old terrain that has produced, among other goodies, the oldest (known) evidence for life on the planet in the form of stromatolites. (Value range G).

Types of BIF

There are two major types of iron formation—Algoma type and Superior type. Both types are recognized by a characteristic signature that is difficult to describe but can be seen in photos of the specimens. Algoma type iron formation is usually massive and localized. It is the type of iron formation usually found in Archean rocks, especially iron-formation associated with greenstone belts and their extensive evidence of volcanism. Superior type iron formation is usually associated with the later Precambrian, the Proterozoic. It is well layered and sometimes exhibits stromatolites where its biogenic origin is more obvious that it is with the Algoma type. Superior type iron formation can also extend over large areas, Algoma Type BIF's by contrast, are usually highly localized.

Superior type banded iron formation, Marquette, Michigan. The Negaunee Formation of northern (upstate) Michigan is mined near Marquette. This is an example of a Superior-type BIF (named for Lake Superior). Observe the distinct bedding consisting of alternating layers of red jasper and blue-grey hematite, a characteristic of Superior-type iron formation.

Another view of the Negaunee Iron Formation outcrop, Marquette, Michigan.

Glacial cobble of BIF. Jasper here is yellow rather than red. The difference between these two forms of BIF being that the iron oxide in the yellow form is limonite or goethite. This form of iron oxide (like rust) contains some chemically combined water molecules. This glacial erratic was found in northern Michigan. It probably came from outcrops to the north in Ontario, possibly from around Espanola, Ontario, where yellow jasper iron formation crops out. (Value range F).

Boulder of BIF from the Sokoman Iron Formation, northern Quebec. This is also an example of a Superior-type iron formation that is not in the Lake Superior region.

Mesabi Range Minnesota iron ore mine. These are some of the largest iron mines on the Canadian Shield. The early Proterozoic (Paleoproterozoic) Biwabik Formation being mined is an excellent example of Superior type BIF. The Biwabik Formation also carries beds of stromatolites—stromatolites like this are known by rockhounds as Mary Ellen jasper.

BIF pebble conglomerate. Thick sequences of pebble conglomerate, often with the pebbles composed of vein quartz, are characteristic of both the Early Proterozoic and the Archean. This Early Proterozoic pebble conglomerate from Western Australia is composed primarily of pebbles of banded iron formation (BIF). The pebbles themselves exhibit a fine layering and granulation characteristic of Algoman type iron formation, that is an Archean type BIF. Mount McGrath Formation of the Wyloo Group, 50 miles ENE of Paraburdoo, Western Australia. (Value range F).

Outcrop of the Gunflint Formation near Nolalu, western Ontario. The Biwabik Formation is the same as the Gunflint Formation in Canada. One of characteristics of the Superior-type iron formation is that it can cover large areas, unlike the older Algoma-type iron formation, which is very localized. The Gunflint Formation is not oxidized here and is predominantly black and grey. In Minnesota's Mesabi Range, the iron has been concentrated by deep weathering over a long period of geologic time and the red of hematite predominates.

Iron Formation pebble conglomerate. Pebble conglomerates often thousands of feet in thickness are found in Early Proterozoic and Archean terrains. Often the pebbles are composed of quartz. In this conglomerate most of the pebbles are composed of banded iron formation. Mount McGrath Formation, Wyloo Group, Western Australia.

Outcrop of Early Proterozoic pebble conglomerate. Thick sequences of pebble conglomerates like this occur associated with early Precambrian terrains. Here most of the pebbles are composed of quartz; second in abundance are pebbles of black chert and banded iron formation like the BIF pebbles in the Australian conglomerates of the previous photos. Vallecito Canyon, LaPlata County, Colorado.

Attigun Pass, Brooks Range, northern Alaska. These grey rocks are Paleozoic in age (Devonian). They are overlain by the iron formation-pebble-containing conglomerate.

Conglomerate of Paleozoic age containing numerous pebbles of iron formation (jasper), Attigun Pass, Brooks Range, northern Alaska. The source of these iron formation pebbles was some Precambrian age rock that no longer outcrops in the area. Iron formation pebbles are distinctive—but determining their age (other than being Precambrian) is difficult.

The Alaskan pipeline before crossing the Books Range at Attigun Pass, which is to the south from this view.

Sphere containing iron formation pebbles from Attigun Pass, Brooks Range, Alaska. The pebble of clast at the center is granite, that to the right is iron formation. (Value range F).

Hot Spring and/or Geyser Deposits

Hot springs represent a type of ancient "fossil" environment that has allowed for the survival of a variety of primitive life forms. They appear to have been much more common and widespread during the early earth than they are today—a scenario which is logical as hot spring (or geothermal activity) is ultimately powered by the heat of radioactive-element-decay and **radioactivity was much more intense in the early earth**. Geothermal springs usually harbor a group of primitive microbes known as archeria (or archaebacteria). These are prokaryotes that have been found to be genetically distinct from other bacteria (the eubacteria); archaebacteria thrive under conditions that are otherwise hostile to other life forms.

Fumerol, Yellowstone Park, Wyoming. These hissing and smelly things almost certainly were widespread during the early earth. It was probably a smelly place! Their geothermal energy comes from the heat of radioactivity, which during the early earth there was more of (as the earth gets older, less radioactive material is present). Three billion years ago, this scene would have looked familiar, without, of course, the dead and alive spruce trees (or any land vegetation for that matter).

Deposits from hot springs and geysers is known as siliceous tufa or geyserite. It accumulates from silica carried in the ascending hot water. Regarding the early earth, such deposits, forming on the surface, are rarely preserved in the rock record, generally being removed by weathering and erosion.

Erupting geyser and hot spring. Associated with such geothermal activity today and (almost certainly) in the Precambrian as well, would have been primitive ecosystems of prokaryotic life. This probably included extremophiles like thermophyllic bacteria living in the upper part of the conduit itself and on the surface chemosynthetic and photosynthetic bacteria as well as related cyanobacteria (blue-green algae). The latter probably lived in moist areas further from the geyser, as can be seen on the left. Sapphire Geyser, Yellowstone Park, Wyoming.

Old Faithful Geyser. Geysers are hot springs with internal plumbing that allows water to become superheated, producing a lot of steam underneath the water column. Sufficient steam pressure building up allows the water column to be ejected with considerable force, as can be seen. Geysers must have been much more prevalent in the early earth for reasons stated previously.

Boiling, muddy water in a hot spring (Paint Pots). An ugly but interesting feature of geothermal areas; they like the other phenomena of the early earth have left little of a "fossil record" as erosion over long time intervals has destroyed them (unless they were in some manner buried and preserved in rock).

Geothermal calcareous tufa deposits. These calcareous tufa deposits from a hot spring probably did not exist during the early earth because of the scarcity of limestone at that time. These deposits formed here because a conduit penetrated a thick bed of limestone, which dissolves in the hot water, and the deposits are precipitated when this water came to the surface to form these rim-stone-dams.

A complex sequence of igneous rocks occur in the geologic center of the Missouri Ozarks. Here occurs "tight-lipped" and shadowy evidence of a remote world of hot springs and geysers. It's a world that included not only the expected sulfurous gases, hissing fumaroles, and geysers of geothermal regions but also a variety of primitive life forms, which produced what are problematic structures. To form the considerable mass of igneous rock found in the geologic center of the Missouri Ozarks took a level of volcanic activity unprecedented by today's standards. This large area of felsic volcanic rock represents volcanic activity of a very violent type. Associated with it should have been a substantial amount of hot spring and geyser activity, which accompanies such volcanic activity. "Fossil" evidence for such activity is often difficult to identify, however. In cutting into massive rhyolitic tuffs, peculiar conduits filled with calcite and hematite were exposed that are considered by the author to be the possible conduits of a geyser or large hot spring.

Looking into a geyser conduit.

Road cut on Highway 67 near Fredericktown, Missouri, where fossil geyser or hot spring conduits were collected while making the cutting.

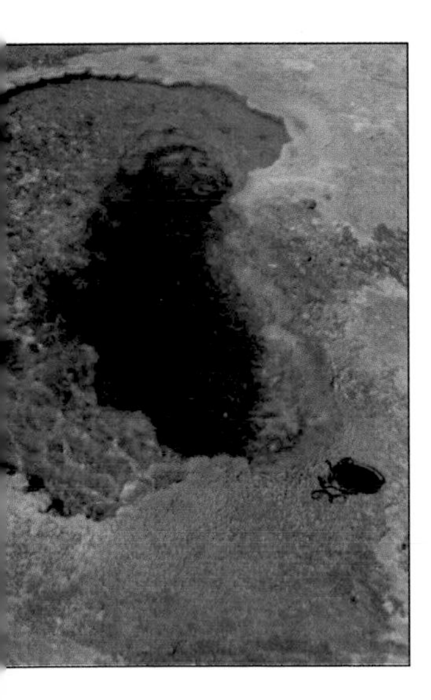

Another geyser conduit, full of hot water with thermophyllic bacteria lining the walls.

Geyser conduit? Excavations in 1.5 billion year old Mesoproterozoic welded tuffs uncovered masses of what appear to be calcite and hematite crystals—crystals which filled geothermal conduits, possibly from a 1.5 billion year old geyser. Slabs sliced from this (presumed) conduit are placed to replicate the way the potential conduit appeared when excavated. Unfortunately, these were completely removed during road construction.

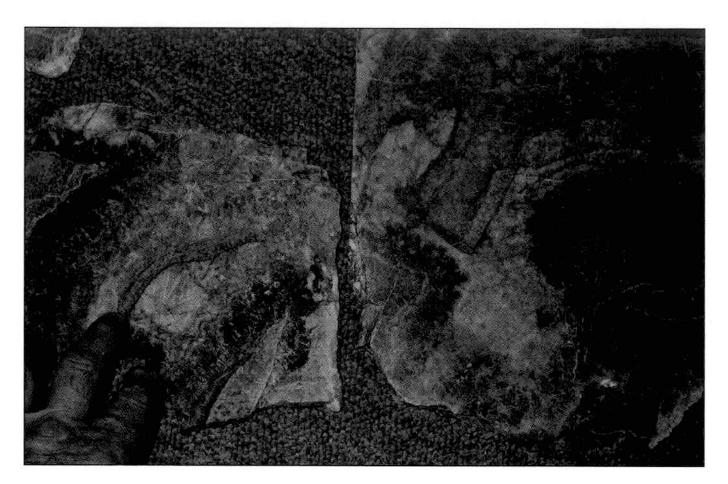

Two cut slabs from geyser or hot spring deposits. Note the presence of oxidization hematite, along with a complex intergrowth of calcite crystals.

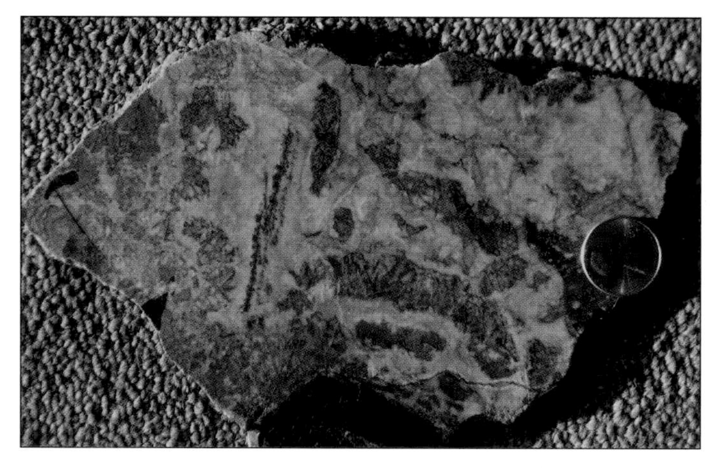

Chunks of hematite-rich material that probably separated from the walls of the geothermal conduit.

Globular masses of calcite filling the geothermal conduit, Fredericktown, Missouri.

Clusters of iron oxide stained calcite crystals from the geothermal conduit.

Red Quartzite of the Early Proterozoic

Precambrian terrains occur that are made up of thick sequences of red or purple quartzite, a rock that is **especially hard**. These rocks extend over a sizeable area and, because of their hardness, may produce good outcrops even in areas of low relief. One of these is the Sioux Quartzite, a reddish rock that extends over a large area where the states of Minnesota, South Dakota, and Iowa come together. The author became familiar early on with this quartzite as boulders and cobbles of it have been widely distributed over the Midwest as glacial erratics—glacial erratics which are both colorful and attractive. Similar in age and appearance is a more purple quartzite known as the Baraboo Quartzite named after the Baraboo Range in southwestern Wisconsin. Like the Sioux Quartzite, the Baraboo occurs widely as glacial erratics east of where the Sioux Quartzite occurs. Both quartzites are early Proterozoic in age, some 1.7 billion years old.

Similar red quartzites are also quite thick and resistant to erosion, occur in the central Rocky Mountains and in New Mexico. These red quartzites occur in China and India as well. They are all quite distinctive and appear to represent part of the "Red Earth" stage of Earth history. With their red hematite pigment, they appear to have something in common with iron formation—the iron in them, like that of iron formation, being in the ferric oxidization state and usually being quite ubiquitous. Red sandstones of later geologic time like that found in the Triassic Period appears to have had an origin different from these Precambrian red quartzites. They are not just red beds subjected to metamorphism.

Sioux Quartzite outcrop. Southwestern Minnesota. This hard, pink quartzite crops out over parts of southwestern Minnesota and eastern South Dakota.

Boulder of Baraboo Quartzite set into a wall. The bands in the boulder are known as Liesegang Bands and are a characteristic of many of these pink and purple quartzites.

Slices of a Baraboo Quartzite glacial erratic with Liesegang Bands.

A string of (heavy) beads made from these red quartzites. Native Americans often utilized quartzite, or meta-mudstone layers associated with it for carvings, including what is known as catlanite. Catlanite was (is) used to make Peace Pipes (Calumets).

Outcrop of massive, early Proterozoic quartzite in the Snowy Range, Wyoming. Like granite, massive quartzite like this, upon being uplifted, does not wear down very rapidly—the rate of uplift exceeding the rare of erosion of this hard rock so that it can form a mountain range (of parts thereof).

Iron Deposits of the Missouri Ozarks

Iron as a major component of Precambrian rocks appears to be the norm in the affairs of Precambrian geology and the early earth. Iron formation and a variety of iron laden sedimentary rocks are the norm—a Precambrian "thing." With this in mind, it should not seem too odd that there are also iron deposits of Precambrian age worldwide that are puzzling as to their origin. One region where such puzzling iron concentrations occur is in the Precambrian rocks of the Missouri Ozarks.

Known since Missouri became a state (one of its counties, Iron County, is named after these peculiar occurrences of iron minerals), mining of iron has been done for over 150 years—much of it in these Precambrian deposits of Iron County. Iron in these deposits occurs primarily as hematite and magnetite. It is associated with igneous rock, felsic igneous rock indicative of a high level of energetic volcanic activity. Iron oxide in these occurrences appears to have been emplaced either in a gaseous or vapor state or as a type of peculiar iron-rich felsic magma. It has been suggested that these large concentrations of iron minerals may represent remobilized iron formation; iron formation that was melted within the crust and then intruded into other igneous rocks as an iron rich magma (or extruded on the surface as iron-rich volcanic ash). Scattered over the Ozark Uplift of Missouri are Precambrian concentrations of iron minerals. These are mainly hematite and magnetite, and many have been mined in the past or are currently being mined.

Beds of hematite-rich volcanic tuff on Pilot Knob, Iron County, Missouri. This bedded hematite rich tuff is believed by some geologists to have been deposited as grains of hematite blown out of a volcano some 1.5 billion years ago.

Head frame of the Pea Ridge Iron Mine, Sullivan, Missouri, 1980.

Iron oxide-rich (hematite) volcanic ash in the Pilot Knob iron deposit.

Calcite crystals from Pea Ridge iron deposit dusted with hematite. (Value range F).

Glossary

Abiotic: Without the presence of life. In reference to the production of organic compounds by means such as demonstrated by the Miller Experiment.

Anoxic Conditions: A chemical environment in which there was an absence of oxygen in the earth's early atmosphere before photosynthetically produced oxygen accumulated to form its "modern" atmosphere.

Auriferous: Gold bearing. In reference to the more frequent occurrence of gold in Archean rocks.

Carbonaceous: Containing fairly large amounts of either carbon or black organic compounds.

Coacervate: A hypothetical "organic soup" of organic chemical compounds existing during the prebiotic, early Earth before the appearance of life. Chemical evolution is proposed to have been the mechanism by which a rudimentary form of life was produced from this accumulation of organic compounds.

Ferric Iron: Iron that has lost three electrons and is present in such minerals as hematite and goethite. Ferric iron compounds generally are relatively insoluble in water.

Ferrous Iron: Iron that has lost two electrons and therefore is in a more reduced state than ferric iron. Ferrous iron compounds are usually water soluble.

The Miller Experiment: An experiment using an electrical spark above a mixture of hydrogen, nitrogen, carbon dioxide, and water vapor, which, in an enclosed flask, produced a mixture of organic compounds including amino acids, the fundamental building-blocks of proteins and living things. This experiment attempted to reproduce conditions of the early earth under which life may have formed. It worked in the sense that it produced some of the fundamental compounds (building-blocks) of life.

Bibliography

Fortey, Richard, 1997. *Life, a Natural History of the First Four Billion Years of Life on Earth.* Vintage Books. Also, *Life, an Unauthorized Biography*, Harper Collins, London) ISBN 0-375-70261-X.

Botryoidal (bubbly) hematite and pyrite on calcite. Pea Ridge Mine. (Value range G).

Calcite and pyrite specimens from Pea Ridge Iron Mine. Mineral specimens from Precambrian iron occurrences in Missouri are usually covered with this fine dusting of hematite. (Value range F, single specimen).

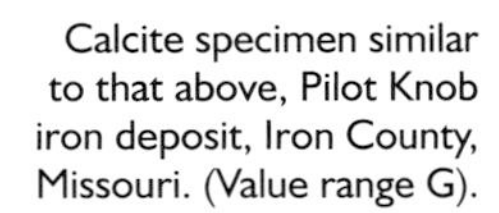

Calcite specimen similar to that above, Pilot Knob iron deposit, Iron County, Missouri. (Value range G).

Hematite "dusting" on calcite crystals, Pea Ridge Mine, Sullivan, Missouri. (Value range G).

Pyrite on calcite with minimal hematite dusting. Pilot Knob Mine, Iron Co., Missouri. (Value range F).

Chapter Six

Life, an Ancient Phenomena on Planet Earth, Early Fossils

A False Alarm—Eozoon Canadense (The Dawn Animal of Canada)

When serious exploration of geology in North America began in the mid-nineteenth century, one of its major quests was to discover fossil evidence for some of Earth's first life. It was realized that North America had greater areas of ancient rocks exposed on its surface than did Europe. This quest for the earliest fossil evidence of life was particularly focused on Canada, as its Canadian Shield had an even greater amount of ancient rocks than did the US. One of the outcomes of this quest was the discovery of Eozoon, or what became known by its scientific (linnean) name as *Eozoon canadense*, or the "Dawn Animal of Canada."

Eozoon canadense was found by J. William Dawson of McGill College (now University), who, a few decades later, in the latter part of the nineteenth century, had this to say about it.

> In addition to this inferential evidence, however, one well-marked animal fossil has at length been found in the Laurentian of Canada, *Eozoon Canadense*, (fig. 7), a gigantic representative of one of the lowest forms of animal life, which the writer had the honour of naming and describing in 1865—its name of "Dawn-animal" having reference to its great antiquity and possible connection with the dawn of life on our planet. In the modern seas, among the multitude of low forms of life with which they swarm, occur some in which the animal matter is a mere jelly, almost without distinct parts or organs, yet unquestionably endowed with life of an animal character. Some of these creatures, the Foraminifera, have the power of secreting at the surface of their bodies a calcareous shell, often divided into numerous chambers, communicating with each other, and with the

J. William Dawson's description of Eozoon in his 1873 work *The Story of the Earth and Man*.

Eozoon turned out to be a false alarm; however, it gave encouragement to other geologists working in ancient rocks of Canada and the US to examine closely other rocks of Precambrian age found at "the bottom of the stack." One of the fruits of this searching somewhat resembled Eozoon; these fossil-like structures were given the general name of stromatolites. Today stromatolites are known to be structures formed in shallow water by the life activities of some of the most primitive of photosynthetic life, the cyanobacteria. The cyanobacteria (or what previously was known as blue-green algae) are related to bacteria, and like bacteria, they consist of cells which lack a cell nucleus (prokaryotic cell type) and belong to one of the five kingdoms of life known as the monera. Monerans were, and are, responsible for forming stromatolites (which are still forming today in restricted areas of hypersaline and fresh water) and were the first forms of life known to exist on the early earth.

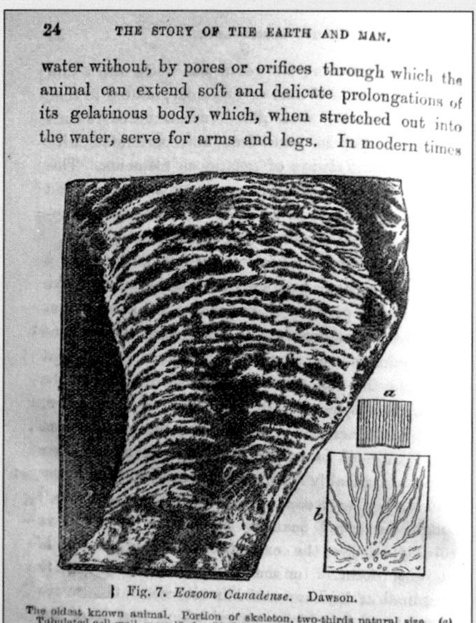

24 THE STORY OF THE EARTH AND MAN.

water without, by pores or orifices through which the animal can extend soft and delicate prolongations of its gelatinous body, which, when stretched out into the water, serve for arms and legs. In modern times

Fig. 7. *Eozoon Canadense*. Dawson.

The oldest known animal. Portion of skeleton, two-thirds natural size. (*a*) Tabulated cell-wall, magnified. (*b*) Portion of canal system, magnified.

these creatures, though extremely abundant in the ocean, are usually small, often microscopic; but in a fossil state there are others of somewhat larger size, though few equalling the Eozoon, which seems to have been a sessile creature, resting on the bottom of

Supposed microstructure in Eozoon (to the right of the Eozoon woodcut), one of a number of properties which supported Dawson's claim as to the biogenicity of Eozoon.

THE EOZOIC AGES. 25

the sea, and covering its gelatinous body with a thin crust of carbonate of lime or limestone, adding to this, as it grew in size, crust after crust, attached to each other by numerous partitions, and perforated with pores for the emission of gelatinous filaments. This continued growth of gelatinous animal matter and carbonate of lime went on from age to age, accumulating great beds of limestone, in some of which the entire form and most minute structures of the creature are preserved, while in other cases the organisms have been broken up, and the limestones are a mere congeries of their fragments. It is a remarkable instance of the permanence of fossils, that in these ancient organisms the minutest pores through which the semi-fluid matter of these humble animals passed, have been preserved in the most delicate perfection. The existence of such creatures supposes that of other organisms, probably microscopic plants, on which they could feed. No traces of these have been observed, though the great quantity of carbon in the beds probably implies the existence of larger seaweeds. No other form of animal has yet been distinctly recognized in the Laurentian limestones, but there are fragments of calcareous matter which may have belonged to organisms distinct from Eozoon. Of life on the Laurentian land we know nothing, unless the great beds of iron ore already referred to may be taken as a proof of land vegetation.*

Dawson's Eozoon Discussion.

Eozoon Canadense. A weathered surface of one of the objects considered by J. W. Dawson to be gigantic foraminifera or rhizopods (as foraminifera were called in the mid-nineteenth century). Charles Darwin's *Origin of Species*, published in 1859, started an earnest quest for fossils that might precede those of the Cambrian Period—Eozoon was one of the first candidates for this.

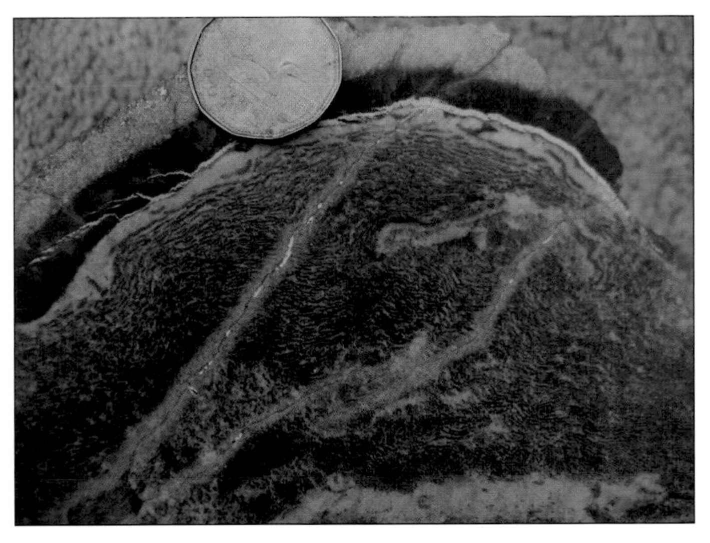

Eozoon Canadense. A cut and polished surface of an Eozoon mass. The author has visited and collected Eozoon in southern Quebec and found that domes like those shown here do resemble parts of a fossil reef, especially when you see the domes in place. Dawson was familiar with Paleozoic fossil coral reefs occurring along the St. Lawrence lowland and in parts of Ontario. The domes of Eozoon appear similar to fossil coral "heads"—one could see why he considered Eozoon to be organic in origin. Note how weathering has infiltrated along cracks.

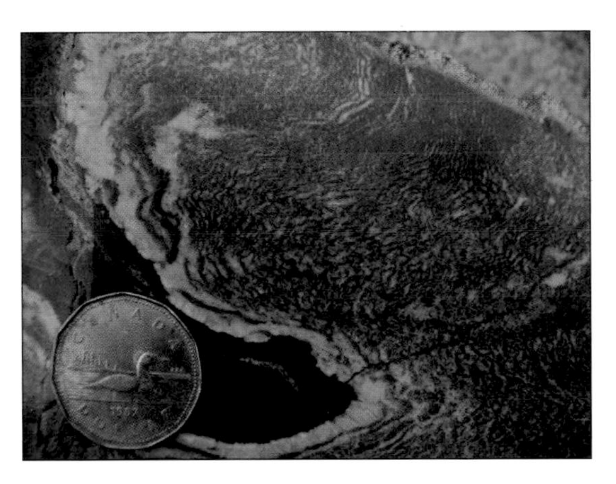

A dome-like extension of Eozoon.

Bonafide fossil foraminifera (or rhizopod). Dawson compared Eozoon to these organisms, which secrete a calcareous test (shell). Some foraminifera in the Eocene Epoch reach 3-5 cm in diameter. This is a specimen of a foraminifera (fusilinid), a normal size for this protist but still gigantic considering that it was a single-celled life form.

Geyser cone. The dark stained area associated with the bottom right of the photo is from a mat of cyanobacteria. This geyser and hot spring is in Yellowstone National Park, Wyoming. These dark surfaces are very slippery and slimy from the cyanobacteria where it grows in the cooler waters. Nearer the geothermal spring outlet are various forms of thermophyllic bacteria and archaebacteria, which thrive in the hot water issuing from the hot spring itself. Without the people and background conifers, this could be a scene from three billion years ago.

Geyser conduit and hot spring. The spherical structures surrounding this pool are made of gyserite—a hydrous (water bearing) form of quartz also known as tufa. They, like stromatolites, are, at least in part, a product of the growth of monerans—thermophyllic bacteria and cyanobacteria. Their interior usually shows a layered stromatolite-like structure.

The dark green and black areas in the foreground are areas of growth of blue green algae (cyanobacteria mats). These hot springs and geysers represent an ecosystem which has graced the earth for at least three billion years. Most of the fossil record of cyanobacteria are found associated with shallow marine environments rather than with hot spring deposits. Hot spring deposits actually are rather rare in the geologic record, usually being eroded away rather than being preserved, not to mention being difficult to recognize in the fossil state.

Geothermal (hot) spring, Yellowstone National Park, Wyoming. The dark areas of this discharge area of the hot spring are formed (primarily) from mats of cyanobacteria. The reddish mat at the bottom left is (in part) from thermophyllic bacteria. These photosynthetic monerans like both the hot water and sunlight.

Erupting geyser, Yellowstone Park. Geothermal regions like Yellowstone represent an ecosystem which has existed from early in the earth's history. Very primitive life forms like thermophyllic bacteria and cyanobacteria thrive today in such a setting, as they did over three billion years ago.

Stromatolites—
The Earliest Direct Evidence for Life

Some of the earth's oldest sedimentary rocks (or their metamorphosed equivalents) contain stromatolites. Many of these are almost identical to stromatolites forming today in shallow water, produced by communities of photosynthetic bacteria and cyanobacteria.

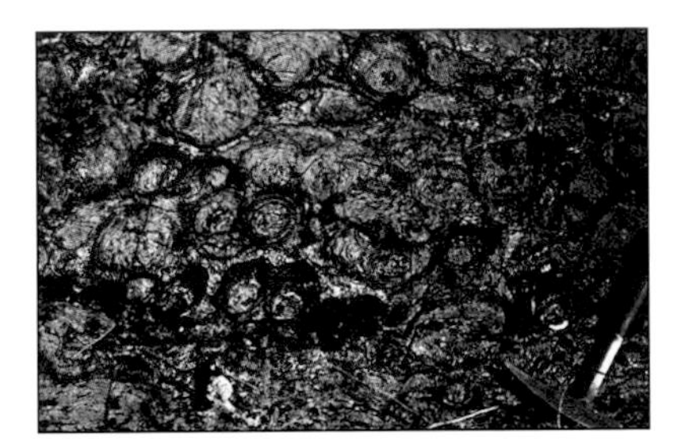

Stromatolites. Part of a 2.3 billion year old stromatolite reef truncated by Pleistocene glaciations in northern Quebec (Ungava) exposed along the Swampy Bay River. Stromatolites form as clustered domes from the growth of cyanobacteria that grew in shallow water on the sea floor.

Close-up large domal stromatolite mass composed of numerous smaller digitate stromatolite "fingers." Swampy Bay River, northern Quebec.

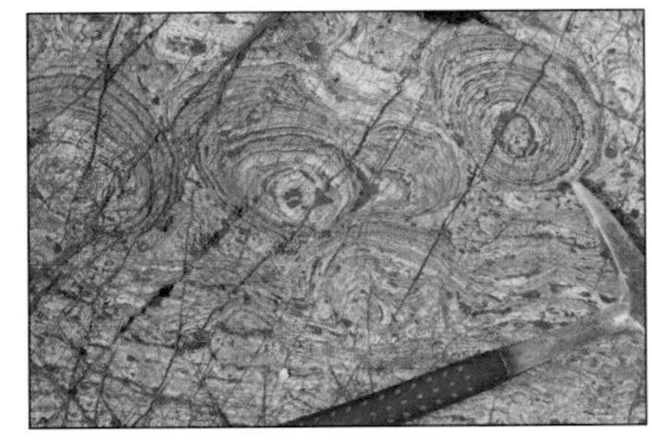

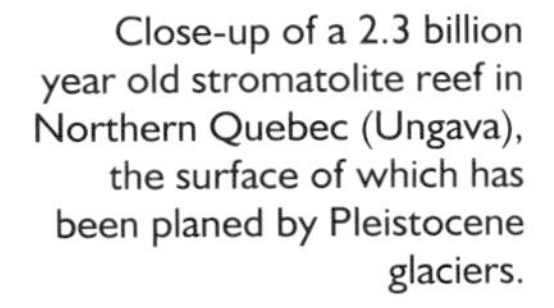

Close-up of a 2.3 billion year old stromatolite reef in Northern Quebec (Ungava), the surface of which has been planed by Pleistocene glaciers.

Large domal stromatolites. These large circular structures are stromatolites. They grew on what was the sea floor as circular mats of cyanobacteria. The Paleozoic sea floor on which they grew has been tilted up at some 60 degrees as the strata containing them has been involved with tectonic activity associated with uplift of the Rocky Mountains Casper Mountain, south of Casper, Wyoming. Note how the stroms are made up of **thin layers of limestone** built-up by photosynthesis of the cyanobacterial mats.

Digital stromatolites. Here numerous "finger like" or digitate stromatolites comprise a dome that was part of a stromatolite reef in the early Proterozoic Labrador Trough or geosyncline of northern Quebec, Canada. Stromatolites come in a variety of forms and shapes. Their form may be due to a variety of variables including water depth, the presence of sea currents, the amount of sunlight present, and its angle of incidence as well as types of monerans present. Of the latter, presumably most were photosynthetic cyanobacteria; however, chemosynthetic and photosynthetic bacteria can also be responsible for some stromatolites. Most stromatolites found as fossils represent colonies of cyanobacteria once growing in shallow marine waters rather than in hot springs.

Close-up of thin layers which compose these stromatolites on Casper Mountain, Wyoming.

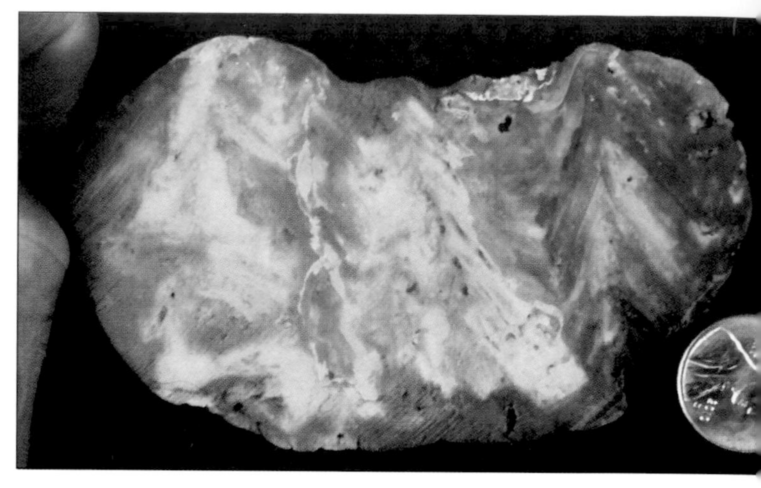

Conophyton sp. Sliced slab showing cone-in-cone structure characteristic of one type of stromatolite. These stromatolites have their concave-upward structure in the form of a cone rather than a concave-upward meniscus. They are sometimes found associated with geothermal springs and may have formed through the presence of thermophyllic bacteria. They are found sporadically in the Ozarks of Missouri associated with some of the major faults of the eastern Ozarks—possibly forming from hot waters that issued along these faults promoting the growth of thermophyllic monerans. St. Genevieve Fault System, Womack, Missouri.

A 2.4 billion year old stromatolite tilted on edge. Medicine Bow Mountains, Wyoming.

Conophyton sp. A large *Conophyton* mass from late Cambrian rocks of the eastern Ozarks, Missouri. These unique stromatolites are found associated with major faults of the eastern Ozarks and may represent the activities of thermophyllic monerans, which thrived from hot water associated with these faults. (Value range F).

Close-up of the previous Medicine Bow Mountains stromatolites.

The top surface of the *Conophyton* mass of the previous photo showing peculiar pattern characteristic of these stromatolites. Specimen from southern Jefferson County, Missouri.

Stromatolite paper weights. A polished cube from a billion year old stromatolite reef from Bolivia, South America. Numerous cubes of this "strom" from Bolivia have been sold as paper weights through the Tucson mineral, fossil, and gem shows. (Value range F).

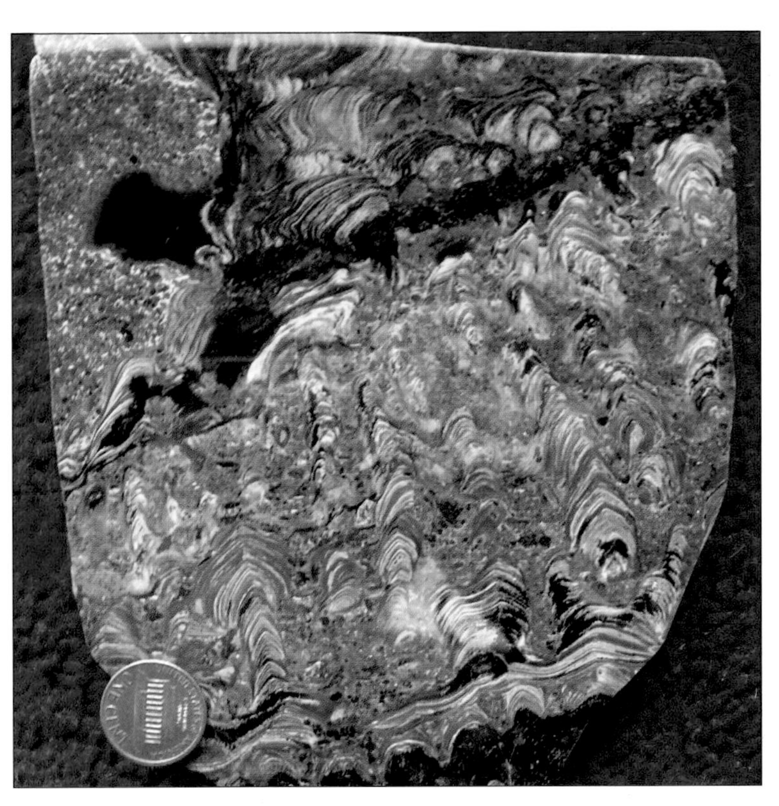

Black, digitate stromatolites (*Collenia undosa*) preserved in black chert from the 2.3 billion year old Biwabik Formation of northern Minnesota. These black chert stromatolites have yielded a biota of small microfossils representing a variety of presumed monerans responsible for the stroms. The Biwabik Formation is the same as the Gunflint Formation in Canada, the formation that has yielded a variety of small (presumed) moneran microfossils.

Digitate stromatolites. "Fingers" in this group of digitate "stroms" have been preserved in quartz rather than with calcium carbonate, the usual material making up stromatolites. This is a thin slab from the Biwabik Formation of northern Minnesota. These stromatolites can show, under high magnification, a biota of microfossils, presumable all prokaryotes but occurring in a surprising variety of shapes and morphologies greater in diversity than found in modern monerans.

Microfossils in Stromatolites

Discovery of microfossils in stromatolites, microbes that appeared responsible for the formation of stromatolites, led to the realization that throughout most of the Precambrian life existed only in the form of single celled, primitive microbes. The absence of fossils other than stromatolites was explained as a consequence of there being a long period of Earth history (over two billion to .7 billion years) when only microbial life existed. If stromatolites found in very ancient rocks (that is rocks of the Archean Era) are taken into account, this period of exclusive microbial existence extended over three billion years of geologic time. (This is a "reading" of the fossil record that didn't take place until the late 1960s). Skepticism of Archean stromatolites still persists today for good reason as these early Precambrian stromatolites, unlike younger ones, have failed to yield convincing microfossils proving their biogenicity.

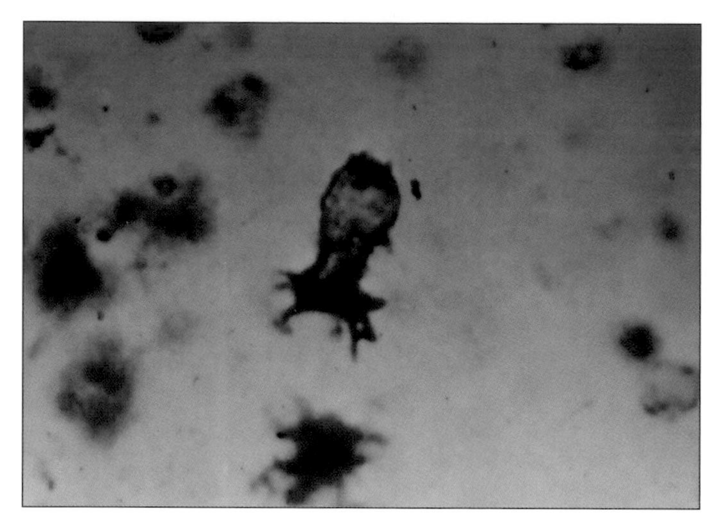

Kakabekia unbilacata. An especially distinctive specimen of this microfossil from stromatolites of the Gunflint Chert of Canada. Small microfossils found in the Gunflint cherts and stromatolites are probably monerans; however, they are morphologically more complex than are most living monerans, which usually are shaped like rods, chains or filaments.

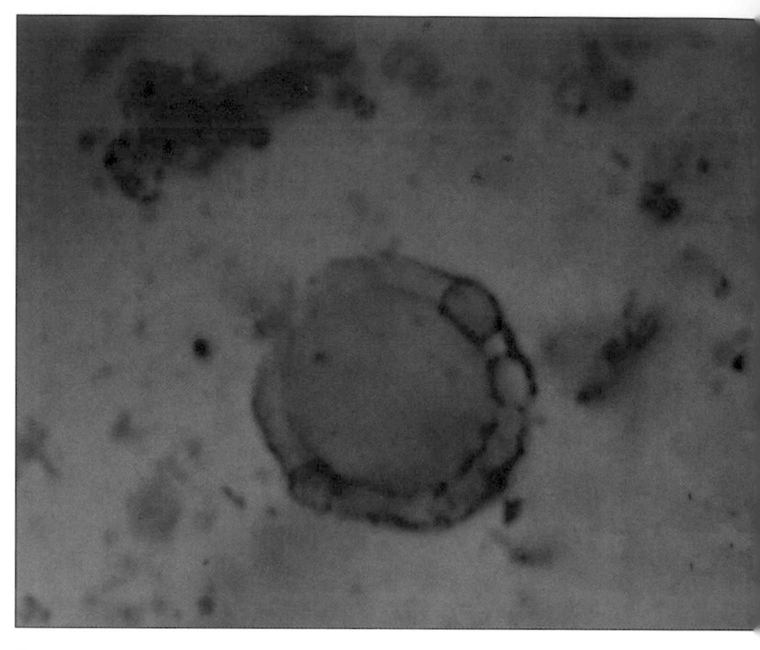

Eosphaera tyleri sp. The most morphologically complex of the Gunflint fossils. *Eosphaera* consists of a larger sphere surrounded by numerous smaller ones. The Gunflint Formation of Canada is known as the Biwabik Formation in Minnesota. Its stromatolites have yielded a variety of puzzling microfossils, some of which are shown here.

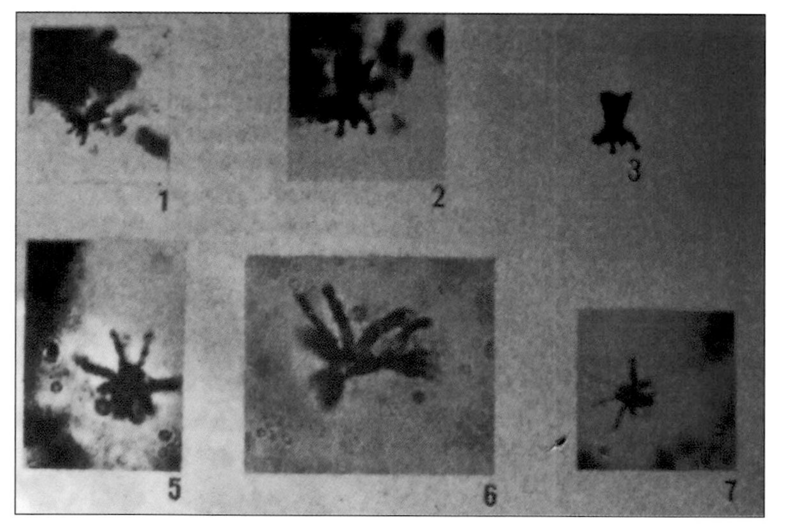

More typical collapsed specimens of *Kakabekia* sp. from the Gunflint Chert.

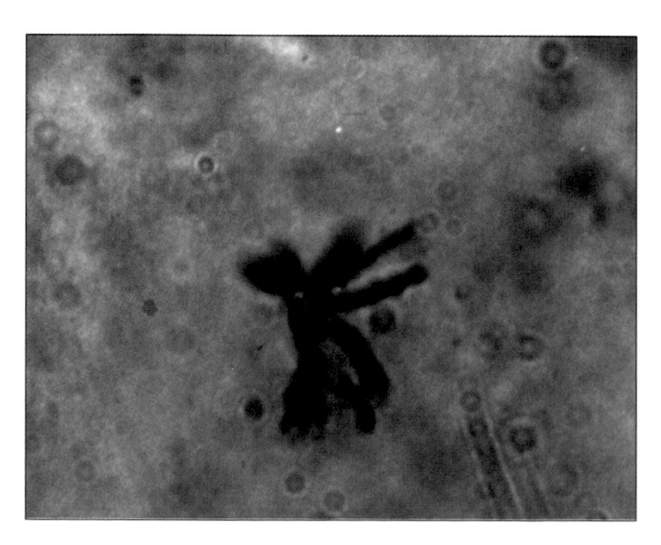

Collapsed *Kakabekia* sp.

Belt Series Stromatolites

The Belt Series is an eight mile thick sequence of mostly sedimentary rocks that crop out in western Montana, northern Idaho, and northward into Alberta and British Columbia as well as in the Yukon Territory. They contain a plethora of nice stromatolites.

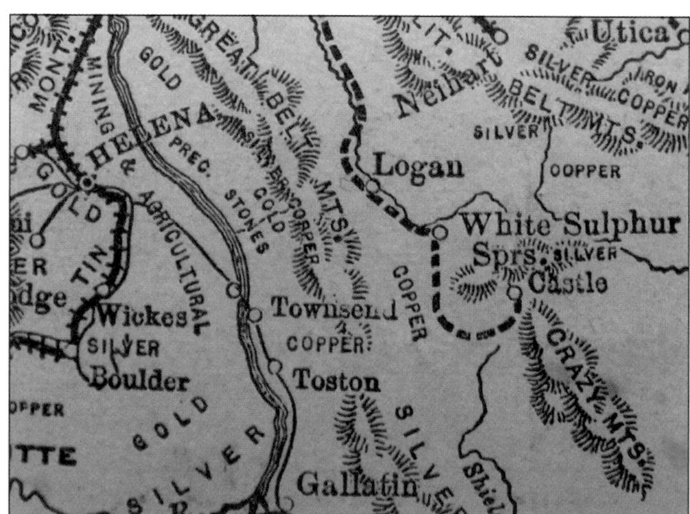

Map of the region around White Sulphur Springs, Montana, and the Big Belt Mountains from which the Belt Series was named. Note that various mineral occurrences are named in this late nineteenth century map.

The Big Belt Mountains of western Montana—the type locality of the mid-Proterozoic Belt Series, a Precambrian rock sequence known for its wealth and diversity of stromatolites. This view is looking west from White Sulphur Springs, Montana. In the ravine just west of where this picture was taken is the locality where Charles Walcott described a variety of odd stromatolites in his "Precambrian Algonkian Algal Flora" paper of 1914.

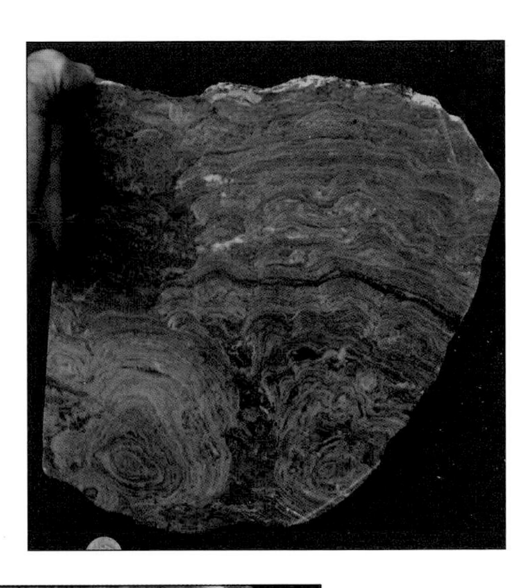

Greysonia sp. A Belt stromatolite that is especially rich in ferric oxide.

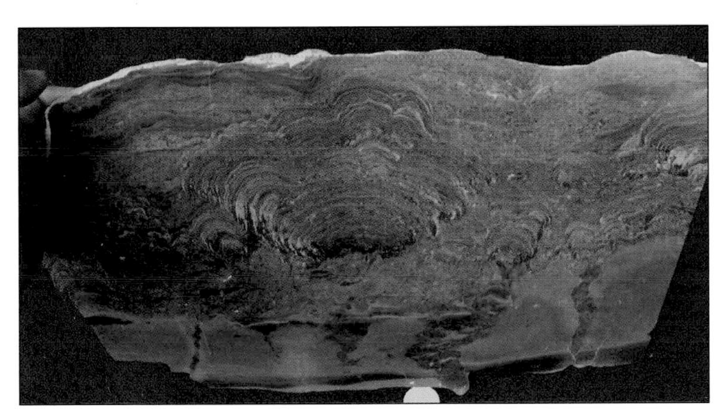

Collenia frequens Walcott. Stromatolite, Belt Series, northwest Montana. Siyeh Limestone, near Kalispal, Montana. This slice is through a well-formed stromatolite colony (note smaller domes adjacent to the large central dome) that shows red as a consequence of containing hematite (ferric oxide). This red pigment came from precipitation of insoluble ferric iron formed from ferrous iron reacting with free oxygen produced by the photosynthesis of the cyanobacteria responsible for the stromatolite. Note the greenish substrate upon which these "stroms" formed. Its green color is from ferrous iron—a more soluble form of iron oxide that is stable in an anaerobic (oxygen free) environment. Such an oxygen free environment included the sediments of the sea floor 1.5 billion years ago (the age of these stromatolites). (Value range E)

Another iron-rich group of various sized stromatolites from the mid-Proterozoic Belt Series (Siyeh Limestone) of northwestern Montana.

Slice from another portion of the same strom as shown in the previous photo. (Value range F).

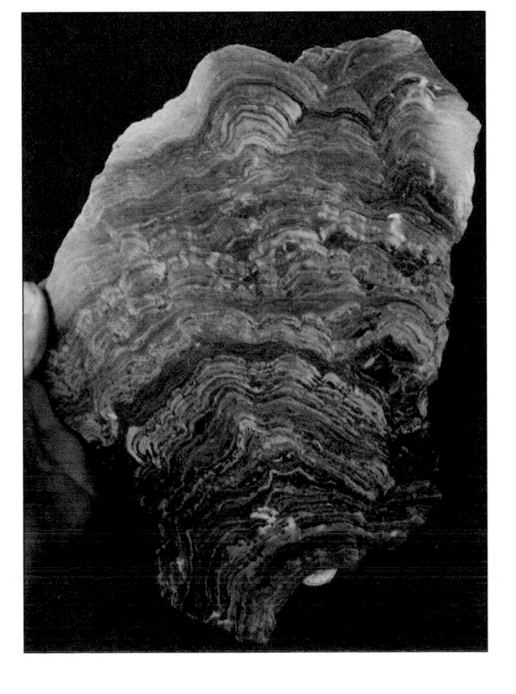

Stromatolites vary considerably from one occurrence to another (as well as varying in morphology as a consequence of water depth, presence of currents, and in what part of geologic time they formed). This variation, however, is normally within relatively strict limits. All of these stromatolites are from the same stratigraphic unit (Siyeh Limestone) and are all rather similar.

Complex group of Siyeh Limestone stromatolites. The white veins are made of calcite, essentially calcite filled cracks produced by tectonic forces associated with uplift and mountain building.

Outcrop of stromatolite-rich limestone of the Newland Formation of the Belt Series just west of White Sulphur Springs, Montana.

Group of low domal stromatolites **in outcrop** of the Newland Limestone, just west of White Sulphur Springs, Montana.

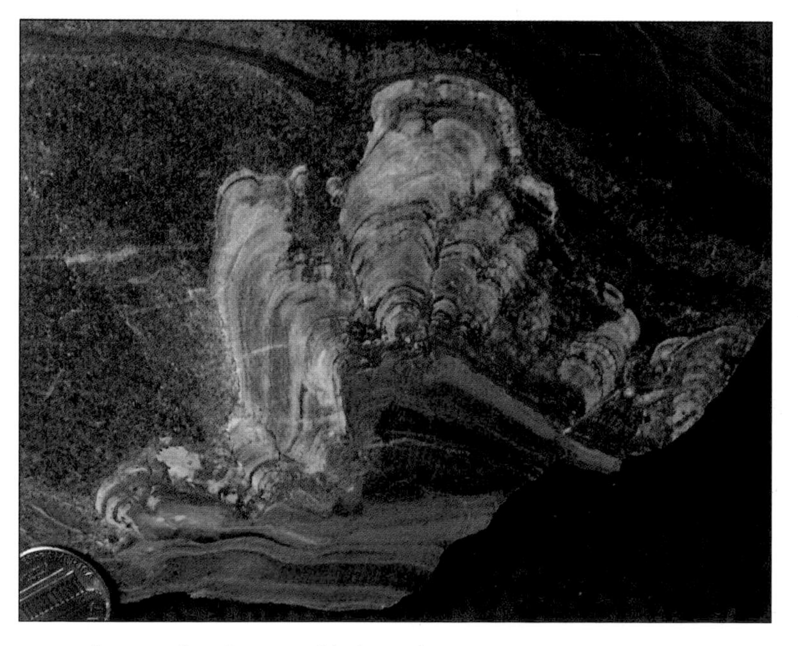

A group of small stromatolitic domes that grew on a fractured portion of the Belt sea floor. (Value range F).

Close-up of the top surface of these low domal stroms from the Newland Limestone west of White Sulphur Springs, Montana.

Another slab of low domal stroms in the Newland Limestone.

Newlandia sp. These peculiar structures come from Australia and are known in the rockhound world as zebra stone. Other than the rock in which they are found (the Australian occurrence being oxidized from weathering in the dry climate), they are identical to those found in Montana. Whatever group of microbes existed to form both these peculiar stromatolites, they were almost certainly of the same type and also of the same geologic age.

Stromatolites sometimes can take on peculiar shapes other than domes and fingers. These stromatolites, given the form genus (name) of *Copperia* by Charles D. Walcott resemble ripple marks. They have been mistaken for fossil ripple marks by geologists not familiar with the wide range of stromatolites. Outcrop of Newland Limestone, 8.5 miles west of White Sulphur Springs, Montana. (Notice the white veins of calcite in the outcrop produced by tectonic forces).

Cryptozoon sp. (*Collenia frequens* Walcott). Stromatolites and stromatolite reefs are well exposed along "Going to the Sun" highway in Glacier National Park, where this outcrop and its stromatolites occur.

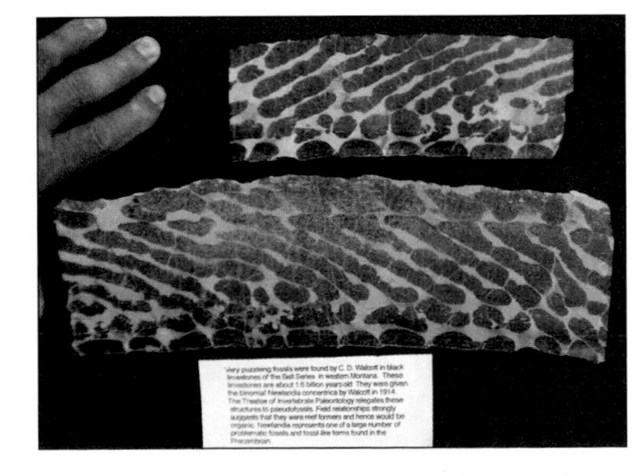

Newlandia sp. This peculiar stromatolite has recently been negated by some geologist-paleontologists to the category of a pseudofossil (false fossil), **which it is not**. It's a peculiar type of stromatolite, which was recognized and named by Charles Walcott as a type of fossil algae (stromatolite). Walcott was not only a good scientist and administrator, but he was also a good field geologist. He recognized these peculiar structures found in the Belt Series for what they are—that is fossils (stromatolites). Lesser field workers more recently have negated Walcott's identifications and placed them as pseudofossils. When one sees them in the field, they obviously are stromatolites—all-be-it odd ones.

Close-up of the previous stromatolite—one of the interesting geologic features of Glacier National Park, Montana.

Mysterious stromatolite? Limestone beds of the Ravalli Group of the Belt Series consists of great quantities of rock with this structure and signature. The pattern is very consistent and may have some sort of biogenic origin. I hesitate to place it as a stromatolite, however. Vast amounts of the limestone in the Ravalli Group exhibit this pattern where it crops out on the border between Montana and northern Idaho.

A (non-stromatolitic) argillite of the Belt Series, western Montana. These slightly metamorphosed and well bedded (or layered) strata are typical of outcrops of the Belt Series in Montana, Idaho, and British Columbia.

A mass of Belt Series strata surrounded by and embedded in diorite, an intrusive igneous rock. This block of Belt strata in Idaho has been emplaced in and surrounded by this hard rock. As a consequence of **deep burial in the crust,** the original rock has been engulfed in magma and "cooked" or metamorphosed. As a consequence of this, the Belt strata in the area where this picture was taken resembles the hard Precambrian rocks so characteristic of the Canadian Shield—rocks that also are full of quartz veins and igneous dikes. Belt outcrops to the east, which bear stromatolites like those illustrated before, were not so deeply buried in the crust as this rock was and look more like Paleozoic rocks. Occurrences like this combine the crystalline rocks of chapter three with fossil bearing strata like that of the Belt Series of western Montana.

Close-up of the bedded meta-mudstone (argillite) mass of the previous photo exposed west of Sand Point, Idaho.

Metamorphosed Belt Series strata exposed west of Sand Point, Idaho. This strata has been intruded by a diorite dike, which formed as a consequence of these rocks being deeply buried in the earth's crust. Belt strata to the east, in Montana, at the eastern edge of the Rocky Mountains, by contrast, have been little affected by either metamorphism or by the intrusion of igneous rocks. They, unlike these rocks, were never deeply buried in the crust.

Close-up of (finely crystalline) diabase dike intruded into metamorphosed mudstone of the Belt Series, northern Idaho.

Same outcrop as in the previous photo, but under different conditions.

Pegmatite dike intruded into metamorphosed Belt Series rocks. Here the highly crystalline, deep seated rocks of chapter three (with their crystals) are found associated with what to the east are fossil bearing rock strata. Here again are two completely different types of geology connected—**hard rock geology** with its crystals and deep crustal origin meeting **soft rock geology** with its fossils and horizontal rock strata. Most geologists never see these two (opposite ends of the spectrum) positioned together.

A Gallery of Proterozoic Stromatolites (Primarily from Australia and China)

The Proterozoic Era (the later half of the Precambrian) was a long span of time during which lived vast quantities of blue green algae or cyanobacteria. This spells out as a sequence of rocks locally blessed with lots of stromatolites—stromatolites often with considerable diversity. Here is some of that diversity—but also variations on the same theme, a theme that played on for hundreds of millions of years.

These different variations form the basis for giving stromatolites names, which are known as "form genera." They are given such binomials even though they were not distinct organisms but rather are a type of trace fossil (as are fossil tracks and trackways), which are also given form genera and species. This strategy has been criticized by some geologists, however, it puts a "handle," or a specific name, for a specific kind of stromatolite (and also enables one to find specific types on the "net", an unanticipated benefit of this strategy). Australian and Russian workers on Precambrian stroms especially have utilized this concept.

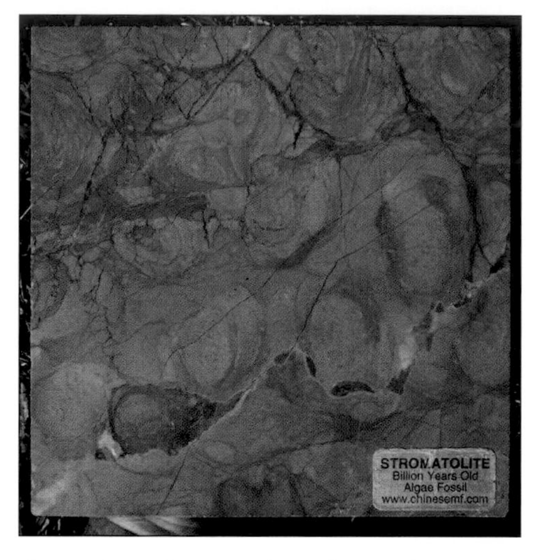

Flower-ring-rock stromatolite with some (post-diagenetic) fractures.

Tabletop made of Chinese "Flower-Ring-Rock." Late Proterozoic Northeast Red Formation. (Value range D).

Chinese "flower-ring-rocks." Reefs of late Proterozoic red limestone (Northeast Red Formation) containing digitate stromatolites are quarried, cut, and polished to be used as a popular decorative stone in China. (Value range F).

Weathered slab of stromatolite reef from which flower-ring-rock stromatolitic marble is cut. (Value range F).

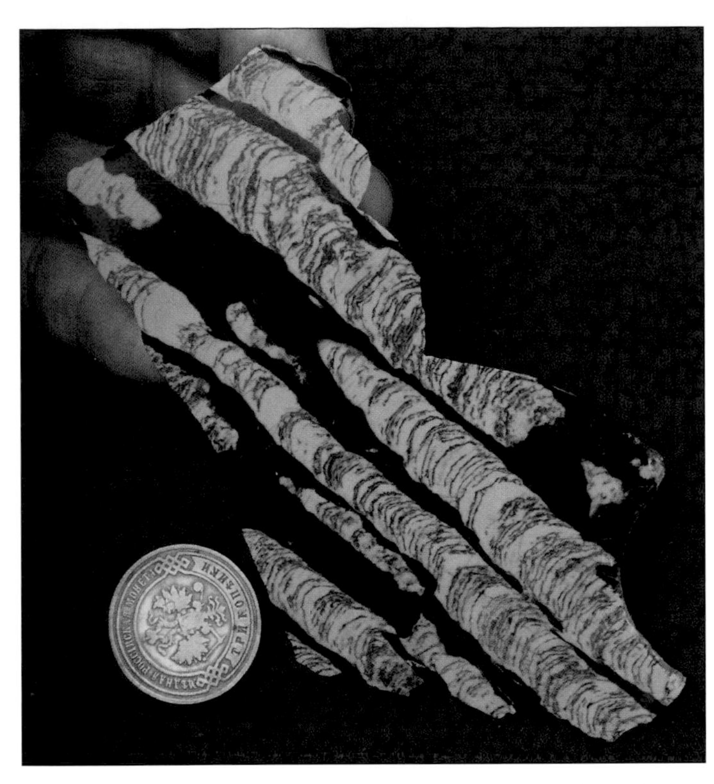

Black and white digitate stromatolite, Russia. These spectacular "stroms" came from the Proterozoic of Russia. Jakkut-Salia, Targo, Russia. (Value range F).

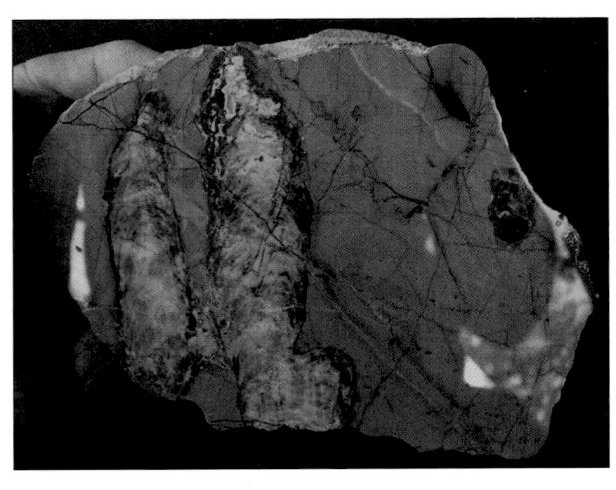

Pilbara perplexa. Duck Creek Dolomite. A digitate stromatolite from 2.4 billion year old rocks (Early Proterozoic or Paleoproterozoic) of the Pilbara Block of Western Australia. (Value range E).

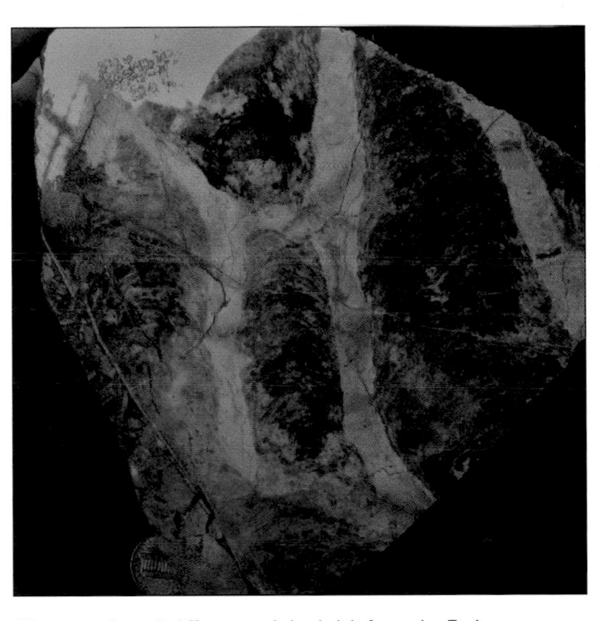

Pilbara perplexa. A different polished slab from the Early Proterozoic Duck Creek Dolomite of Western Australia, near Paraburdoo, Western Australia. (Value range F).

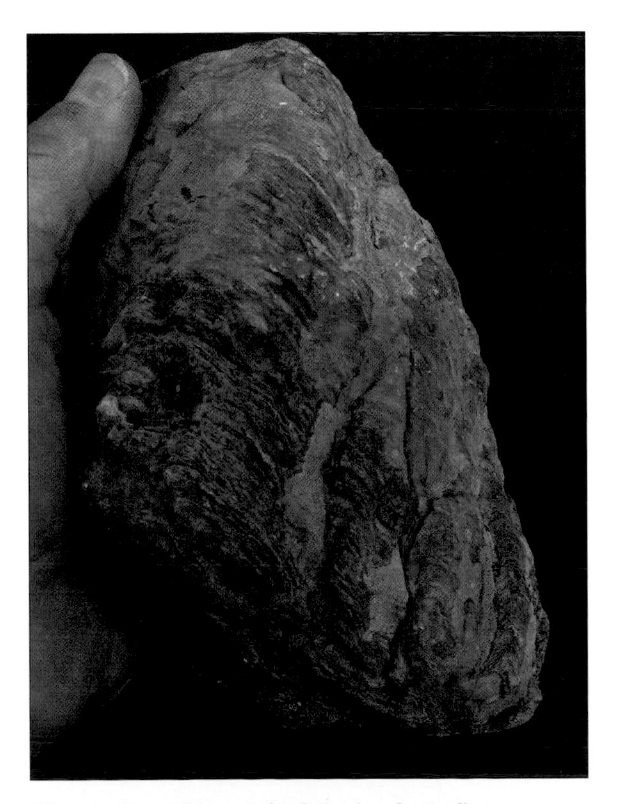

Pilbara perplexa. **This and the following Australian stromatolites represent form genera and species designated by Australian workers on "stroms."** Specimen with naturally weathered surface. Duck Creek Dolomite, Western Australia. (Value range E).

"*Cryptozoon*" sp. A slice through a large dome of this "generic" stromatolite. Irreguelly Formation, Western Australia. (Value range E).

Pilbara perplexa, Duck Creek. A portion of a colony of this early (Lower Proterozoic, Paleoproterozoic) stromatolite from younger strata of the Pilbara Block of Western Australia. (Value range F).

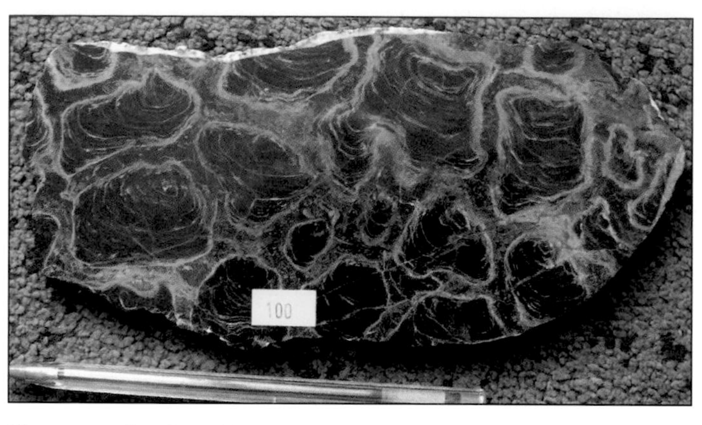

Minjaria pontifera. Bitter Springs Formation, Alice Springs, Northern Territories, Australia. (Value range F).

Kulparia alicia. Bitter Springs Formation, Northern Territories, Australia. (Value range F)

Asperia ashburtonia. Bitter Springs Formation, Alice Springs, Northern Territory, Australia. Value range F).

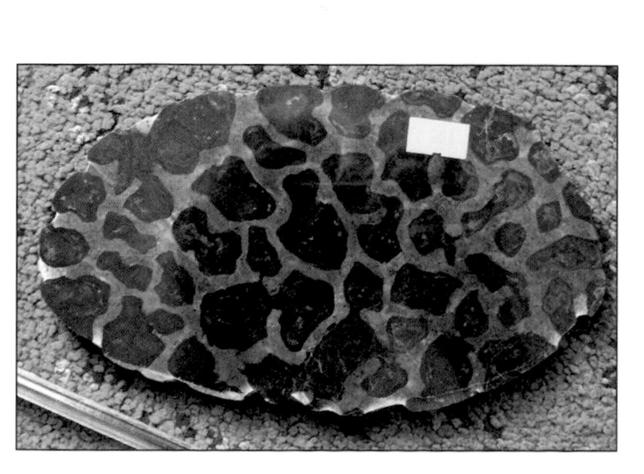

Boxonia pertaknurra Walter. Bitter Springs Formation, near Alice Springs, Northern Territory, Australia. (Value range F).

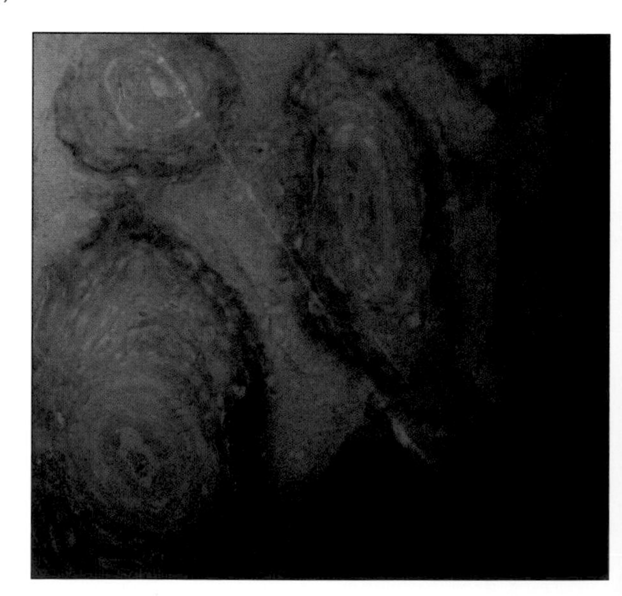

Kulparia Alicia, close-up. Bitter Springs Formation, Northern Territories. This is similar to the Chinese "Flower-Ring-Rock" stromatolites.

Alcheringa narrina. *Alcheringa* is a distinctive laminar stromatolite from the Proterozoic of Australia. Lower Proterozoic. Tumbiana Formation. Pilbara, Western Australia. (Value range F).

Acaciella australica. Polished slab, Bitter Springs Formation, Northern Territiories, Australia. (Value range F).

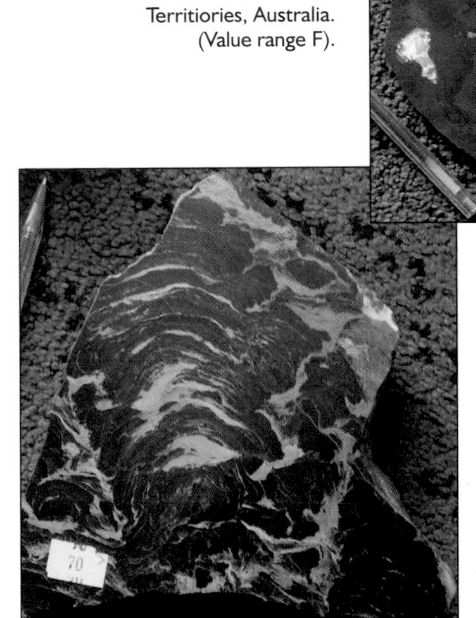

Acaciella australica. Specimen composed of a large dome. Bitter Springs Formation. Near Alice Springs, Northern Territory, Australia. (Value range F).

Ozarkcollenia laminata. Ketcherside tuffs, Mid-Proterozoic. Cuthbertson Mountain, Iron County, Missouri. This laminar stromatolite found in 1.5 b.y. volcanic tuff from Missouri, closely resembles *Alcheringa* from Australia. (Value range F).

Azurite stromatolites? These azurite buttons are believed by some Australian geologists to be either small stromatolites, which were replaced by copper minerals, or possibly they were stromatolites produced directly by chemosynthetic bacteria. A colony of chemosynthetic bacteria can utilize, as metabolic energy, the chemical energy produced from the oxidization of heavy metals like copper. Some sulfide ore deposits are believed to have had a similar origin as does the metallic sulfides found with "black smokers" associated with deep sea geothermal vents and occurring at deep sea trenches. As geothermal activity appears to have been more common in the Precambrian, considering radioactive decay as the heat source for such activity, such a hypothesis as to the origin of these azurite "buttons" seem plausible. These "azurite stroms" were first brought to the attention of the northern hemisphere geologic community by the late Allen Graffham of Geologic Enterprises, Ardmore, Oklahoma. (Value range F).

Inzeria intia. Peculiar cup-cake-like stromatolites from the Bitter Springs Formation, Alice Springs area, Northern Territory, Australia. (Value range F for group).

Close-up of azurite stromatolites? Gardiner Range, Alice Springs, Northern Territories.

Close-up of bottom specimen of azurite stroms(?) shown in the previous photo.

Archean Stromatolites

Stromatolites are rare in the Archean, the earliest part of the Precambrian. Those which do occur are associated with greenstone belts. One of the reasons "stroms" are rare in the Archean is that most of the earth's early oceans were deep. Few regions of shallow water existed, as shallow waters of the oceans are associated with continental shelves and there were few or no continents during the Archean.

Archean stromatolites push evidence for life on the earth prior to three billion years. Proof of their biogenicity is of considerable interest and importance because they place the origin of life at very early in earth's geologic history. Stromatolites indicate that the appearance of life on Earth was an early and ancient phenomenon. Early appearance of life might also give some insight as to whether life was introduced to the earth from elsewhere in the galaxy (panspermia) or evolved here through chemical evolution not long after the violent and hostile conditions of the formative stages of the Hadean Era ended. Archean

Gigantic late Archean stromatolites formerly exposed in the Steep Rock Lake iron mine near Atikokan, western Ontario. These large domes of mega-stromatolites are composed of ferruginous (iron bearing) limestone and are some of the largest and oldest "stroms" known. They are now covered by the waters of Steep Rock Lake and are no longer exposed as the mine is no longer in operation.

Two of the mega-domal stromatolites formerly exposed at the Steep Rock Lake Iron Mine, Atikokan, Ontario.

stromatolites also are of considerable interest in the field of exobiology. Life may also have appeared on (or was introduced to) Mars in its early history, which at this time, appears to have been more similar to Earth. In other words, this early appearance of life, as documented by stromatolites, makes them of considerable interest to exobiology; but, this requires a **clear documentation as to their biogenic origin**. This can be difficult as early Precambrian rocks are often highly metamorphosed and in other ways are peculiar. Sometimes these rocks are even made up of minerals usually not associated with fossils.

Small domal stromatolites that compose the stromatolite layers making up the large "stroms" shown in the previous photo.

Value and "Hype" Regarding Archean Stroms

Stromatolites are relatively common fossils. Stroms in Archean rocks, however, are seen a lot less frequently than are younger ones. There are only about seventeen Archean stromatolite occurrences known worldwide. Current interest in exobiology, coupled with the fact that these stromatolites represent the earliest evidence of life, has "jacked up" the price of them on the collectors market. This might be feared to adversely affect their availability to science; however, stromatolites, as is the case with many other fossils, occur in strata that can, if diligently and carefully collected, yield a lot of fossils, especially considering that slabs or thin slices of stromatolites generally show their structure as well as do large chunks.

Laminar stromatolites also compose the huge domes of the Archean stromatolites formerly exposed in the Steep Rock Lake Mine. (Value range F).

These "wavy" 3-5 inch thick limestone beds were formerly exposed in the now water-filled Steep Rock Lake Iron Mine. They are not from folding or tectonism, but rather are sections through the mega-stromatolites shown above. These huge stroms are in turn composed of much smaller laminar and digitate stromatolites shown in the following photos.

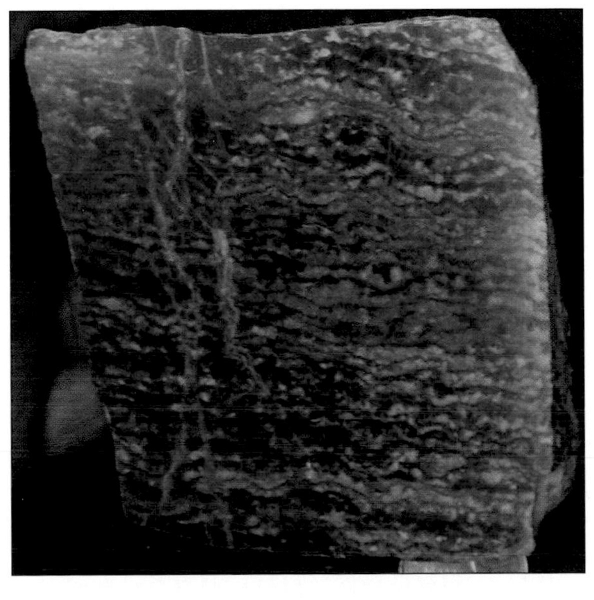

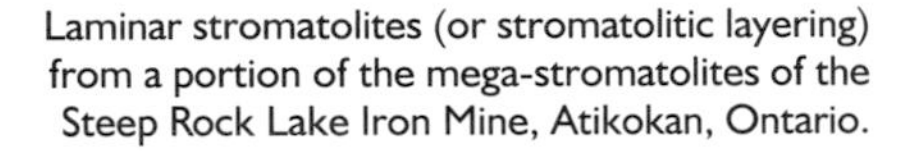

Laminar stromatolites (or stromatolitic layering) from a portion of the mega-stromatolites of the Steep Rock Lake Iron Mine, Atikokan, Ontario.

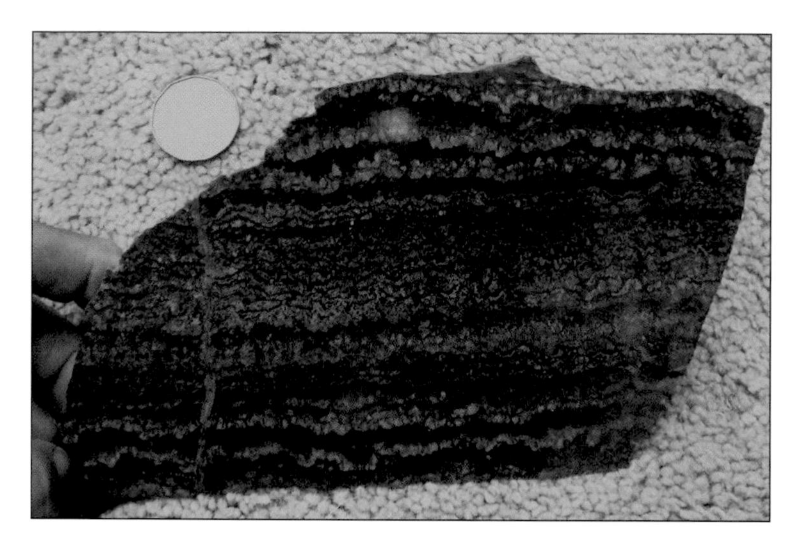

Slab of laminar stroms (with small, indistinct domes), Steep Rock Lake Iron Mine, Atikokan, Ontario.

Ferroan dolomite crystals (tan) and quartz crystals (white) in cavities (vugs) found throughout the mega-stromatolites of Steep Rock Lake Mine. (Value range F).

Group of similar crystals as above from former Steep Rock Iron Mine.

Kussiella sp. A stromatolite from a large glacial erratic found in northeast Missouri and believed to be from the same late Archean sequence as the Steep Rock Lake stromatolites. The late Archean belt of strata that yields the Steep Rock Lake stroms (as well as a pseudofossil or dubiofossil known as Atikokania) probably extends into northern Minnesota. There, however, most outcrops in that area are now heavily covered with other glacial drift. The large boulder from which these black stromatolites came was associated with glacial erratics suggestive of an Archean terrain on the southern part of the Canadian Shield, probably from northern Minnesota.

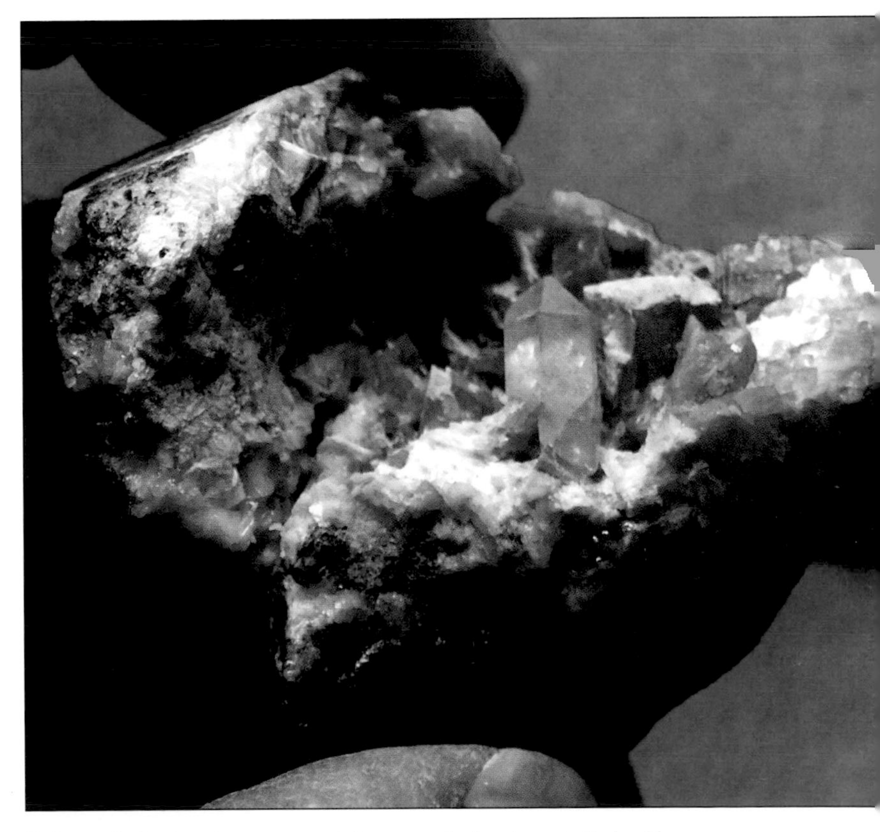

Dolomite (yellow) and quartz crystals from a vug in the stromatolite boulder mentioned above. Vugs like this are characteristic of large stromatolites. Some of the fine crystals, as found in the Missouri lead belts, are also associated with vugs related to large stromatolites.

Stromatolite(?) from a mid-Archean calcareous iron formation, Wawa, Ontario. Mining of iron-rich dolomite beds associated with an Archean Greenstone Belt near Wawa, Ontario, encountered an occurrence of these in the ore zone. They appear to be strongly metamorphosed stromatolites.

Algal limestone from the Belingewe Greenstone belt, Zimbabwe, Africa. Limestone associated with stromatolites, including those of the Archean, have a distinctive signature, which is hard to describe, but, as has been said regarding pornography, "It's hard to define but I can really tell it when I see it." The same holds for many limestones associated with stromatolites.

Linked domal stromatolite (Bulawayan stromatolites), Belingewe Greenstone Belt, Zimbabwe. Archean stromatolites are usually black; rarely are they of the reddish or brown colors commonly associated with Proterozoic stroms. The predominance of black stromatolites is a consequence of the reducing or anoxic environment under which they grew and which prevailed during the Archean. Free oxygen during the Archean did not exist in the atmosphere. Bulawayan stromatolites from Zimbabwe were for decades the earth's oldest fossils and evidence for the first life. Rocks of the Belingewe Greenstone Belt age date at 3.3 billion years.

Small domal stromatolite in algal limestone (marble) of the Belingewe Greenstone Belt, Zimbabwe, Africa.

Laminar stromatolite—Strelly Pool Chert, Pilbara Block, Western Australia. ***Currently these are the oldest known stromatolites and therefore are the earliest (minimally questioned) evidence for life.*** They occur in Archean rocks, which make up the center (oldest part) of the Pilbara Block. This is a greenstone belt exposing mid- and early Archean rocks, most of which are not too severely metamorphosed. Strelly Pool stromatolites and the limestone in which they are found are not black or as dark colored as is the case with most Archean stromatolites. This lighter color and the more oxidized condition of this laminar strom, and the rocks in which it is associated, may be a consequence of geologically recent weathering of the rocks in the dry climate of Western Australia. (*Courtesy of Curvin Metzler*).

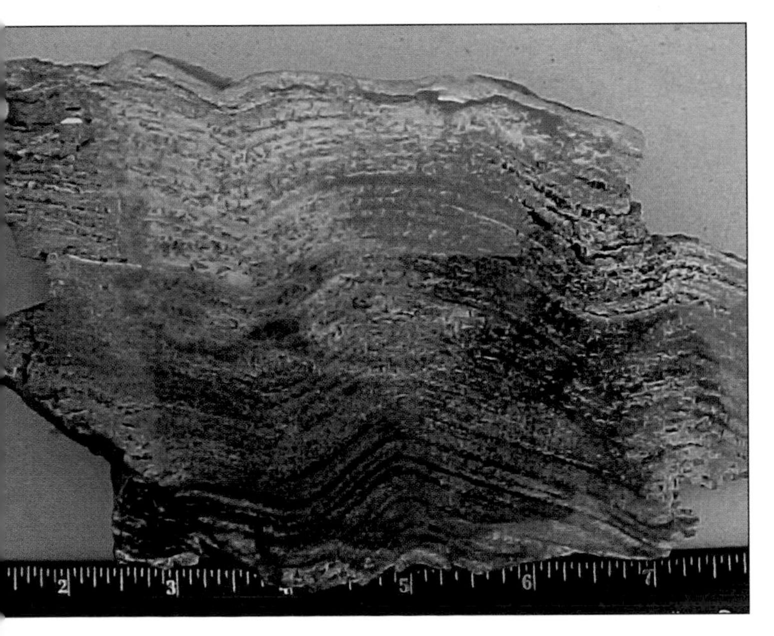

Strelly Pool Laminar Chert stromatolites showing less oxidization than that of the previous photo. These stromatolites (probably) were probably originally black from the reducing conditions of the earth's early seas. Most Archean rocks and associated structures (like stromatolites) show that they were formed under reducing conditions in which little or no free oxygen was present.

Ediacaran Organisms or Vendozoans

Arguably these are the strangest fossils of any found in the earth's rock strata. Ediacaran fossils not only are strange looking, but even stranger is what they may represent. Originally found in southwest Africa (Namibia), then later in the Ediacara Hills of eastern Australia, they have also been found in England, Russia, Newfoundland, North Carolina, and elsewhere, they represent a real **paleontological enigma**. For decades, they were considered to be various forms of lower invertebrates like soft corals (sea pens), jellyfish, and early arthropods and echinoderms in the case of the Australian occurrence in the Ediacara Hills, the occurrence for which they were named (Glaessner, 1961). Ediacaran fossils (or Vendozoans after the Vendian, the youngest portion of the Russian Precambrian) are found in the youngest of Precambrian rocks. They are found in what are known as Neoproterozoic rocks, strata that lie just below rock yielding the earliest fossils of the Cambrian Period. Ediacarian organisms are considered by some paleontologists as a type of "experimental" life form, which went extinct before the Cambrian Period. They are envisioned as being of a leathery consistency, possibly containing symbiotic bacteria, which may have been their source of metabolic energy. Others have suggested that they were a type of marine lichen. Lichens today have a symbiotic arrangement between fungi and photosynthetic algae, which includes the cyanobacteria. There is some indication that these strange organisms may have formed parts of moneran matts, the organisms growing as a part of the matt itself.

Sea cliff outcrops of the late Precambrian Conception Group contain numerous impressions of vendozoans exposed at Mistaken Point, Cape Race, southeastern Newfoundland.

Seashore exposures of the late Precambrian Conception Group, Cape Race, Newfoundland. These are the most profuse occurrences of vendozoans known and are also the geologically oldest.

Bedding surface of meta-mudstone of the late Precambrian Conception Group with a profusion of spindle-shaped vendozoans (rangeomorphs). The brown growths are living (modern) sea weeds as this unweathered exposure is within the surf zone.

Somewhat weathered bedding surface of meta-mudstones of the Conception Group with numerous rangeomorph vendozoans. Concentration of these peculiar fossils is greater here than at any other (known) vendozoan locality.

Large vendozoan below the hammer. Some Ediacarian vendozoans are quite large. The distinct impressions that they form led to suggestions that they were, when living, stiff and leathery (like an air mattress). Inside this "air mattress" were photosynthetic symbiotic monerans. The Conception Group vendozoans appear to have lived in deeper water than those found in eastern Australia.

Medusiform vendozoan. These were first considered to be the impressions of jellyfish (which they do resemble).

Bedding surface with numerous vendozoans.

Close-up of large rangeomorph below the hammer.

Spindle-shaped vendozoan, with margin to the left and right of a medusiform vendozoan.

Bedding surface covered with vendozoans. Conception Group, Mistaken Point, Newfoundland.

Algal mat. An impression of an algal mat in sandstone. Algal mats, a mass of small colonies of cyanobacteria, occurred with some frequency in the Proterozoic as well as in the Cambrian and Lower part of the Ordovician Period. Ediacarian vendozoans have been suggested as being parts of such mats.

Dickinsonia sp. Rawnsley (formerly Pound Quartzite). Ediacara Hills, Australia.

Bibliography

Fedokin, Mikhail A., James G, Gehling, Kathleen Gray, Guy Narbonne and Patricia Vickers-Rich, 2007. *The Rise of Animals. Evolution and Diversification of the Animal Kingdom.* The John Hopkins University Press, Baltimore.

Glaessner, M. F., 1961. Precambrian Animals. *Scientific American.* Vol. 204, pp 72-78.

Knoll, Andrew., 2003. *Life on a Young Planet. The First Three Billion years of Evolution.* Princeton University Press.

LaFlamme M., Guy Narbonne and M. M. Anderson, 2004. Morphometric analysis of the Ediacarian frond *Charniodiscus* from the Mistaken Point Formation, Newfoundland. *Journal of Paleontology*, Vol. 78, No. 5, pg 827-837.

Schopf, J. W. and Walter, M. R., 1993. "Archean Microfossils. New Evidence of Ancient Microbes" in J. W. Schopf, Ed., *Earths Earliest Biosphere. Its Origin and Evolution,* p. 214-139. Princeton University Press, Princeton N J.

Stinchcomb. B. L., 1970. The Oldest Known Fossils, Part I, *Earth Science* Vol. 23, No. 3, May June 4. Part II, *Earth Science* Vol. 23 No. 4, July Aug.

Walcott, Charles D., 1914. *Precambrian Algonkian Algal Flora.* Smithsonian Miscellaneous Publications.

Nemiana sp. Cotter Formation, southern Missouri. An early Paleozoic example (?) of a vendozoan found in Russia.

Chapter Seven

Greenstone Belts and Phenomena

Greenstone

Greenstone is a type of rock usually derived from basalt, which has changed from the pressure associated with deep burial. It is a greenish colored mafic rock. However, not all greenstones are green; some examples are still black. In the process of being drug deep into the earth's crust, these ancient rocks were preserved from weathering over vast time spans by being protected by deep burial. Ancient rocks on Earth, unlike those of planets like Mars and Luna (which have sluggish tectonics), have to be buried deep in the crust for a long time to be protected from weathering and other surfacial geologic processes essentially absent from the other terrestrial planets. Greenstones can also consist of hard, black rock that originally may have been dirty sand or siltstone containing a large amount of mafic material. This is a type of sedimentary rock known as greywacke. Greywacke was a common rock of the early earth. **Greenstone belts make up the oldest known rock sequences on the earth.**

Greenstone. High pressure converts mafic silicate rocks such as augite (a pyroxene) and hornblende (an amphibole) into green silicate minerals such as epidote or diopside. Gas bubbles (vesicules) in the original basalt will be filled with different colored minerals in ancient greenstones.

Ely Greenstone, northern Minnesota. One of the best known and documented Archean greenstone belts occurs in northern Minnesota and western Ontario. The Ely Greenstone is named for Ely, Minnesota, near which it crops out.

Mafic rock in greenstone belts is not always green. Often it is still black as was the original basalt. Rocks in greenstone belts are always very hard and they defy erosion quite well. In this photo, the older, black rock has been intruded by younger red granite, west central Manitoba, Canada.

Metamorphosed Archean greywacke shot full of quartz veins. Greywacke is "dirty" sandstone made up of small fragments of basalt as a major component. Greywacke was produced during the Archean in large quantities and when metamorphosed (as are most greenstones) can be difficult to distinguish from basalt. The bedding (or layering) seen here indicates its sedimentary origin. Nonacho Lake, southern NWT Canada.

Outcrop of Archean basalt—part of a greenstone belt. Greenstones are not always green, originally being basalt or greywacke, which is black. Metamorphism from deep burial in the crust doesn't always change the ferromagnesian (mafic) minerals to epidote, the green mineral of greenstone.

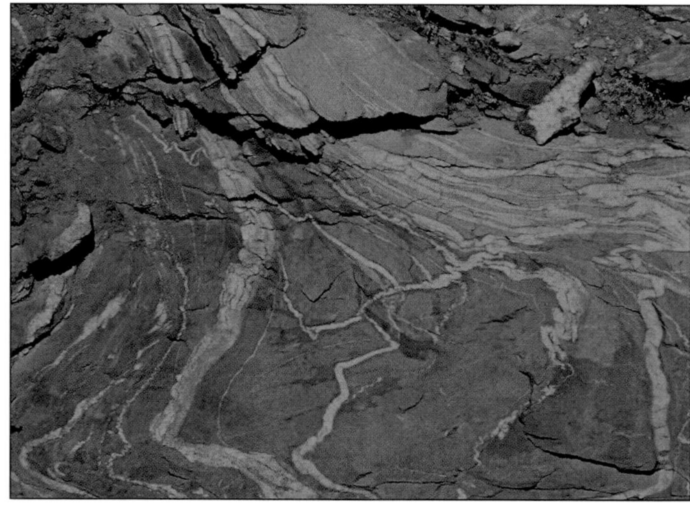

Archean meta-greywacke, Mt. Zirkel, northern Colorado. This metamorphic rock (with quartz veins) was originally a type of dirty sandstone or siltstone.

Outcrop of metamorphosed iron formation, which forms part of a greenstone belt. Iron formation (or BIF banded iron formation) may be a biogenic sedimentary rock related to the early re-dox history of the earth.

Archean rocks usually are quite hard and resistant to erosion. Here an Archean basalt (or gabbro) in western Montana (Stillwater Complex) forms a waterfall as erosion has not been able to appreciably wear down this hard rock.

Copper bearing Archean (?) greenstone from Michigan's Keweenawan Peninsula. Native copper occurs in this area associated with basalt and greenstone in what is a puzzling and anomalous concentration of this native element.

Gold in Greenstone Belts

Greenstone belts represent one of the earth's major sources of gold. Gold mineralization sometimes is found as auriferous (gold bearing) pyrite, as well as with the gold associated with numerous quartz veins that intersect the greenstones. Sometimes this gold in quartz is visible to the naked eye, but more likely it is not.

Surface excavations, Homestake Mine, Lead, South Dakota. Black slates of Archean age are gold bearing in the Black Hills of South Dakota. Placer gold found in streams during the 1876 Black Hills Gold Rush was derived from these rocks, which eventually became the Homestake Gold Mine.

Black, graphitic slate. Metamorphosed sedimentary rock like this graphite bearing slate makes up portions of greenstone belts. Bedding (and cleavage) in these ancient sediments is almost always vertical, as seen here. Near Lake Nipigon, Ontario.

Quartz mass in a hard, meta-basalt near Timmins, Ontario. One of the largest gold producing areas on the Canadian Shield occurs in Archean greenstones at Timmins, Ontario. Gold can occur both as native gold in quartz like this or as gold bearing (auriferous) pyrite.

Gold bearing quartz, Timmins, Ontario. Breaking such quartz chunks occurs in the hope of finding small veins or flakes of the yellow metal. This, however, is not a common occurrence, even in gold producing regions like this.

Native gold in quartz, Timmins, Ontario. (Value range E).

Quartz from a quartz vein with native gold, Colorado. Gold is the bright, yellow component of the vein quartz. The duller yellow mineral is pyrite. (Value range E).

Diabase Porphyry

Not necessarily related to greenstone belts (other than they are both mafic), these intrusive igneous rocks are especially characteristic of the Proterozoic Era of the Precambrian. They represent a very distinctive and attractive rock characteristic of the Proterozoic earth.

Outcrop of diabase porphyry. This distinctive rock appears to be especially characteristic of the mid-Precambrian. It consists of a hard, mafic matrix (groundmass) into which are dispersed well formed sodium plagioclase (sodium feldspar) crystals, Madison County, Missouri.

Polished slab of diabase porphyry from the previously shown outcrop in Missouri.

Glossary

Porphyry. An igneous rock made up of two distinctively different size crystals. In reference to diabase porphyry.

Re-dox. In reference to the oxidization state of iron in the early earth. Iron in a more reduced state is the more soluble ferrous iron, in a more oxidized state is ferric iron. The state of iron (and other multi-valent elements) in reference to their different oxidation states is its re-dox potential.

Slice of diabase porphyry in which the feldspar crystals are replaced by potassium feldspar. This rock, known as Devonite, is known only from the Precambrian of Missouri. (Value range G).

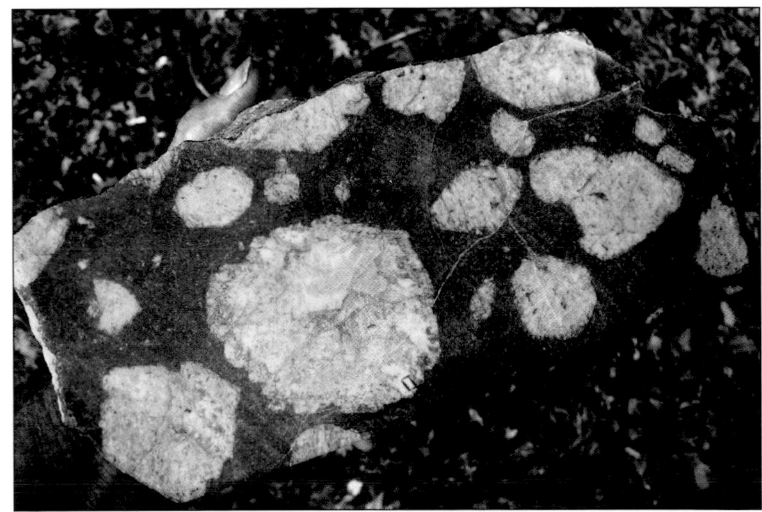

Diabase porphyry in which the feldspar crystals are less well formed than in previous examples. Proterozoic of the northern Big Horn Mountains, Wyoming. (Value range G).

Chapter 8
Marble and Quartz

A Different Look at Early Minerals

Space exploration has subtly changed the way geologists look at Earth! Space exploration has given definitive information regarding the composition, mineralogy, and even the geologic history of the planets and their satellites (along with meteorites). All of this ultimately points out just how unique Earth really is. Standing high in this terrestrial uniqueness are rocks made of carbonate minerals. Such minerals are composed of limestone and dolomite (dolostone), along with their ancient metamorphic derivative, **marble**.

Stated in another way, Earth is a **unique planet** (even disregarding life) as **nothing in the solar system comes close to its geology**. Closely associated with life itself, **carbonate rocks and carbonate minerals appear to be a geological phenomenon especially characteristic of Earth**.

Marble and Quartz

It's been hypothesized that much of the mineral diversity of the earth's crust would not exist were it not for the early appearance of life and the biosphere. This especially seems to be the case with limestone and dolostone, rocks often associated with biogenic activity and the precursor of the metamorphic rock **marble**. The minerals of this chapter are associated with and formed in an environment that originally was limestone—calcium carbonate formed from a marine setting. Limestone and dolostone buried deep within the earth's crust and then subjected to immense pressures produced marble. Limestone is (and was usually) deposited in shallow ocean water. Some limestones can be quite ancient, going back as far as the Archean Era. When beds of limestone become deeply buried in the earth's crust, perhaps five to ten miles, they become "cooked" by heat and hot water. Under these conditions, limestone crystallizes and changes to marble. Such metamorphosed masses of carbonate rock are especially characteristic of the Precambrian. This is because over great spans of time, being drug deep into the crust, they have also had enough time for whatever covered and protected them to have been removed.

Marble outcrop. This outcrop of 1.3 billion year old Grenville Marble in Ontario contains a rock **unique to planet Earth**! Although having no obvious trace to any connection with life, **without Earth's biosphere such a rock could not have formed**. Marble is changed or "cooked" (metamorphosed) limestone and limestone (and dolostone) is a rock unique to Earth. Grenville Marble, Quebec.

Carbonatites

Masses of carbonate minerals not only make up beds of marble, but can also compose the peculiar igneous rock known as carbonatite. Carbonatites (probably) had their origin as beds of limestone or dolomite melted under high pressure and temperature to form a carbonate magma. The parent material of either marble or carbonatite was formed originally from chemical precipitation influenced by the biosphere. That biosphere, in the early earth, was exclusively associated with the biogeochemical activity of monerans. Thus these peculiar rocks could not occur in any part of the solar system other than the earth (assuming that satellites like Europa or Titan lack any type of biosphere).

Marble egg. This egg is made of true marble and true marble* is calcite recrystallized through metamorphism. The black specks are graphite, probably from carbonaceous (organic) material present in the original limestone before metamorphism took place. Most true marble is geologically old; it takes hundreds of millions (or billions) of years to emplace limestone deep enough into the earth's crust and then for whatever material covered it to be eroded away. Marble, like granite, sometimes is said to have formed at the **roots of mountains** and then the mountains eroded over vast spans of time to bring the marble onto Earth's surface. *(*Much of the so-called marble used in building actually is limestone, which is capable of taking a polish, rather than true marble).*

The Grenville Series

Over the southeastern portion of the Canadian Shield in Quebec, eastern Ontario, moving southward into New York (Adirondack Mountains) and western New Jersey is a highly metamorphosed sequence of rocks sometimes containing large quantities of marble. This ancient series of distinctive rocks is known to geologists as the Grenville Series. This rock series (named after Grenville, Quebec) is also somewhat unique for the Canadian Shield in that it contains such large amounts of marble. The presence of such significant quantities of marble in Precambrian rocks, from the author's experience, is fairly unusual. No other part of the Shield that he has seen or is aware of contains such quantities, especially considering the Grenville marbles located near the Quebec-Ontario border. Grenville marble is distinctive in being highly crystalline—often being made of large, interlocking crystals, much larger than those found in most other marbles. These crystalline marbles have been intruded by igneous rocks in dikes of granite, gabbro, and nepheline syenite. At an interface between the marble and the igneous rock, known as a skarn zone, many interesting, rare, and attractive minerals can be found. Intensely deformed and metamorphosed by being buried deep within the crust, impurities in the original limestone on becoming intensely crystallized create these suites of interesting and rare minerals. Attractive minerals which occur in the host rock made up of interlocking calcite crystals, the coarsely crystalline marble. A few other ancient limestone sequences similar to this occur around the globe and these also can be the host for interesting and sometimes rare minerals, including some precious gem minerals.

Greenish apatite crystal in coarsely crystalline marble of the Grenville Series, Otter Lake, Quebec. These gemmy apatite crystals are found in highly crystalline marble of the Grenville Series and have been widely distributed among mineral collectors and rockhounds. (Value range G).

Apatite in pink-orange calcite crystals that is really a super-crystallized marble of the Grenville Series. Otter Lake, Quebec, Canada. (Value range G).

Apatite crystals (large) with tourmaline in pink-orange calcite. Otter Lake, Quebec. These minerals would not exist on the earth were it not for the presence of oceans and the presence of a biosphere. This is because of the existence of the common rock known as limestone. Limestone is calcium carbonate. It forms in bodies of water either from physical-chemical conditions that allow for its precipitation (often with the aid of cyanobacteria) from shells and tests of sea life or from other factors (too technical to go into in a work of this sort) that rely on the presence of water and a biosphere. It is sometimes stated that limestone (and hence marble) originates from ancient fossil shells. This statement, in the case of the Grenville Marble, is untrue, in part because of its great age. However for younger limestone (and some marble), this is a reasonable statement. (Value range G).

Clear apatite crystals from Grenville Marble. Tory Hill, eastern Ontario. Grenville marble is found primarily in Canada, equally divided between the provinces of Quebec (French) and Ontario (English). When loyalists, after the American Revolution, settled in the Canadian wilderness, they preferred and utilized these marbles for both construction and decorative stone. (Value range G).

Opaque apatite crystals, Bancroft, Ontario, Canada. A large variety of attractive crystals and minerals come from highly crystallized marble in the area of Bancroft, Ontario. (Value range G).

Opaque apatite crystals, Bancroft, Ontario. Crystals from this locality have been widely distributed among rockhounds. Grenville Series in the states do not contain the quantity of marble that they do in eastern Ontario and Quebec. The odd mineral occurrences at Franklin, New Jersey, being an exception. (Value range F for group).

Zincite and calcite, Franklin, New Jersey. Specimens from this locality have been popular with rockhounds for years, especially as most of the (often unusual) minerals from these old zinc mines are highly fluorescent under a black light. (Value range G).

Dravite. This brown form of tourmaline is from a skarn zone associated with Grenville metamorphic rocks in central New Jersey. (Value range F).

Skarn Zone Minerals

A skarn zone represents an occurrence where an igneous rock (which was once molten magma) was injected or intruded into limestone or marble. When this happens, the molten rock reacts with the calcite (and its impurities) to produce what often are unusual and beautiful minerals. Some of the most diverse mineral assemblages known are associated with skarn zones.

Schorl. Black tourmaline like this usually is associated with pegmatites (see Chap. 3); this schorl comes from mica schist in China where it is (probably) associated with a skarn zone. Similar crystals like this are found in schists of the Grenville Series associated with skarn zones. A large number of these crystals came through the Tucson, Arizona, show in 2009. (Value range E for group).

Garnet var. Grossularite, York River skarn zone, Bancroft, Ontario. Where marble has been intruded by igneous rock, the contact zone (skarn zone) can be the site of unusual minerals, including a variety of garnets. (Value range G).

Titanite. Bear Lake Diggins, Tory Hill, Ontario. Titanium minerals, especially crystals of large size, like this one, usually are associated with geologically unique phenomena like skarn zones. (Value range G).

Sodalite (lapis lazuli). Bancroft, Ontario. This mineral is associated with a skarn zone where Grenville Marble has been intruded by a nepheline syenite dike. This is one of numerous attractive minerals from this region of eastern Ontario. (Value range G).

Lazurite (or Lapis Lazuli) and Sodalite

Lazurite is a mineral that has been known since ancient times as lapis lazuli. It is mentioned in the Bible's Old Testament for both its color and desirability. This attractive mineral occurs as a component of highly metamorphosed limestone that originally contained some sulfur. Sulfur is present in the parent limestone in the form of finely disseminated pyrite. Chemically known as a sulfosalt, lazurite usually occurs as a contact mineral where a bed of marble is in contact with another rock or has been intruded by igneous rock in a skarn zone. Lazurite also may contain small crystals of pyrite, which adds to its interest as a gem material. Lazurite and sodalite occur in the Grenville marble at a number of localities in eastern Ontario, the best known locality being around the town of Bancroft. Most of the lazurite found on the mineral market today comes from Afghanistan where Precambrian marble has been intruded by igneous dikes similar to the occurrences in Ontario. Sodalite and lazulite are chemically similar, with sodalite containing some chloride while lazurite contains more sulfur, which occurs in place of the chloride. Both minerals can be a beautiful blue and this is their major attraction.

Lapis Lazuli. Sodalite and lapis (or lapis lazuli) are unusual colored silicate minerals. The blue color is developed as a consequence of chemical combined sulfur and chloride with sodalite having more chlorine and lapis more sulfur. This polished chunk of lapis is from its best known locality, Koksha Valley, Afghanistan. This blue mineral has been known from what is believed to be the same locality since ancient times. It is even mentioned in the Old Testament as coming from what is believed to be this locality. (Value range C for large polished specimen).

Close-up of a large, polished cobble of lapis lazuli, Badakhshan Province, Koksha Valley, Afghanistan. (Value range C).

Polished pebbles of lapis from Afghanistan. (Value range F for group).

Lapis Lazuli or lazulite crystal. A nearly perfect crystal of this odd mineral (and quite rare) in white marble from a skarn zone associated with marble. Badakhshan Province, Koksha Valley, Afghanistan. (Value range A).

Polished cobble of lapis lazuli, Koksha Valley, Afghanistan. (Value range C).

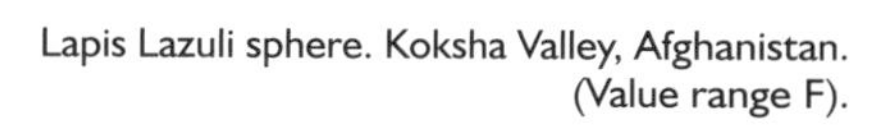

Lapis Lazuli sphere. Koksha Valley, Afghanistan. (Value range F).

Sodalite. Sodalite is similar to lazulite, containing chlorine (as chloride) rather than sulfur. Like lazulite, sodalite occurs associated with marble in skarn zones. Poços de Caldas, Minas Gerais, Brazil. (Value range G).

Ruby corundum. High pressure produced during deep seated metamorphism produced a white marble similar to that of the previous photo. Impurities in the marble, on metamorphism, gather together and produce beautiful crystals like this and the previous spinel. This ruby formed from aluminum bearing minerals (probably clay minerals in the original limestone). An Phu Marble Mine, Lue Yen District, Yen Pai Province, Viet Nam. (Value range F).

Red spinel crystals in white marble. High grade metamorphism can produce attractive and valuable crystals, some of which, like this are gem stones. The beautiful ruby-red spinel crystals occur in a white statuary marble. Mogok, Myanmar (Burma). (Value range F).

Spinel. Somewhat similar to the previous spinel crystals in white marble, these probably formed in a similar geologic environment but later the marble was weathered and dissolved. The spinel crystals accumulating in the remaining regolite formed from the weathering of the marble. (Value range F).

Biogenic(?) Phenomena

Metamorphism of carbonate rocks can produce some interesting phenomena besides minerals. One of these, especially when considering the history of paleontology, is Eozoon canadense. Discussed in detail in Chapter Six, Eozoon is a product of metamorphism associated with impure marble of the Grenville Series of Canada. The Eozoon structure can be a complex arrangement of minerals that suggest the structural complexity of an organism. For this reason it was considered in the late nineteenth century to be fossil evidence for the earth's first animal life, coming on the "heels" of Charles Darwin's "evolutionary theory," a theory which required that some sort of fossil animal life predate the relatively complex life found in the Cambrian Period. Eozoon filled this bill! However, Eozoon is now considered a pseudofossil; it was a false alarm!

Eozoon canadense. A peculiarity associated with high levels of metamorphism of impure limestone can be the formation of structures that resemble fossils. Eozoon was a candidate for the earth's earliest life. It occurs as coral-like masses in marble in what resemble reefs in the Grenville Series of Quebec and eastern Ontario. Considered in the nineteenth century to be a type of primitive animal (a large rhizopod or foraminifera), Eozoon turned out to be a false alarm—a pseudofossil or false fossil. Its structure of alternating layers of calcite and serpentine are suggestive of some type of fossil, like a coral or stromatolite. The tan colored mineral in this slab is a granular form of diopside, a calcium, magnesium silicate. Lac Charlebois, southern Quebec. (Value range E).

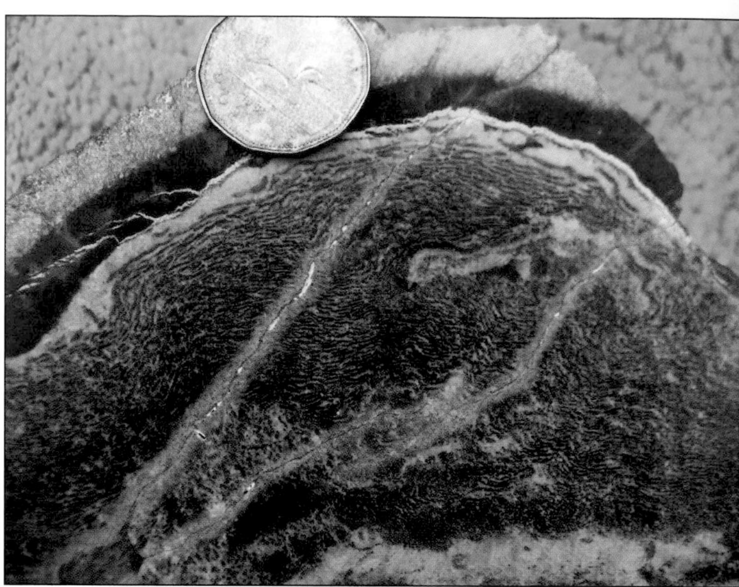

Eozoon canadense. Grenville Marble, Lac Charlebois, Quebec. Charles Darwin's method of evolution by natural selection required that there be fossils of life prior to the Cambrian Period. Eozoon was a "false alarm" which tried to verify this. Stromatolites and Ediacaran Vendozoans (Chapter six), discovered in the twentieth century, would supply some of the answers to this mystery. (But only some!)

Eozoon canadense (top) and Archaeospherina (bottom). A slice through an Eozoon mass is at the top. At the bottom are small globular masses considered by late nineteenth century proponents of Eozoon to be the reproductive "buds" of Eozoon and given the linnean name of *Archaeospherina*. Both of these structures were produced during intense levels of metamorphism when limestone is converted to crystalline marble under deep burial and high hydrostatic pressure within the earth's crust.

Blue diopside. Diopside associated with Eozoon is a light tan, blue diopside is more unusual and rarer. Lake Baikal, northern part, Severobaikalsk, eastern Russia. (Value range F).

Epidote crystals. Epidote is a calcium aluminum silicate. It forms from metamorphism of iron and calcium bearing minerals, which can include the calcium carbonate of fossil shells and stromatolites. These crystals formed from calcium-carbonate-containing-rock metamorphosed to produce rocks of the Grenville Series. This specimen and many other epidote specimens on the mineral market come from Grenville age rocks in the Adirondack Mountains of northern New York State where this slab originated. (Value range F).

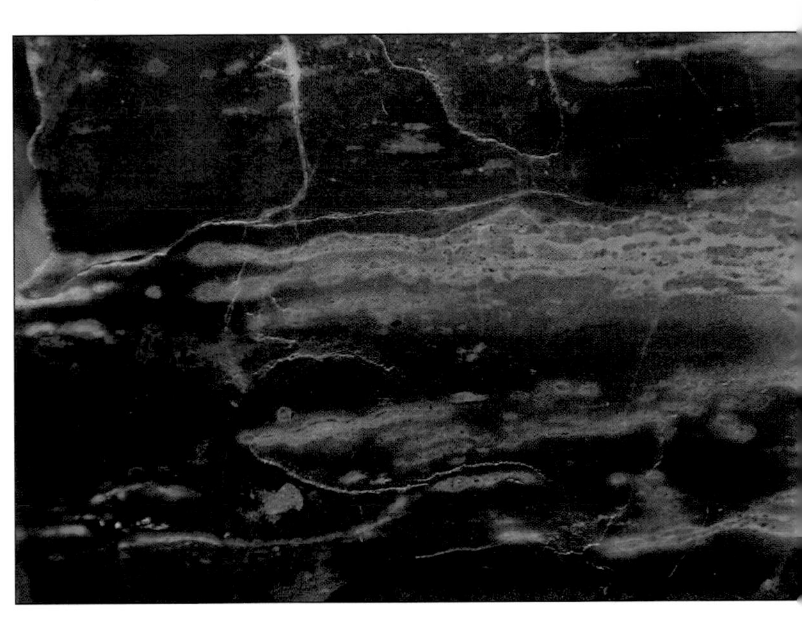

Ozarkcollenia preserved in epidote. What are believed to be the small domes of *Ozarkcollenia* have been converted to epidote from metamorphism of what originally were small stromatolite domes composed of calcite. This occurrence is in a volcanic ash (tuff) of about the same age as that of the previously shown specimen—both are from the Mid-Proterozoic of the Missouri Ozarks. Cope Hollow Formation, Reynolds County, Missouri.

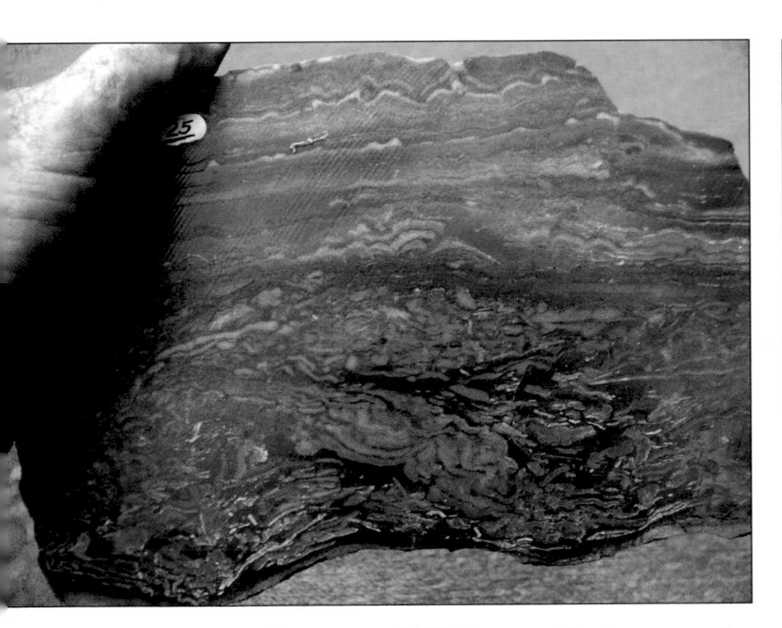

Ozarkcollenia laminata. This is a type of 1.5 billion year old laminar stromatolite from the Missouri Ozarks. It is similar to *Alcheringa* sp. of Australia (see Chapter six), both occur in volcanic tuff. The small pink domes were produced by colonies of cyanobacteria that grew in a large lake or a shallow sea. The stromatolites themselves are composed of calcite with a small amount of manganese carbonate. (Value range F).

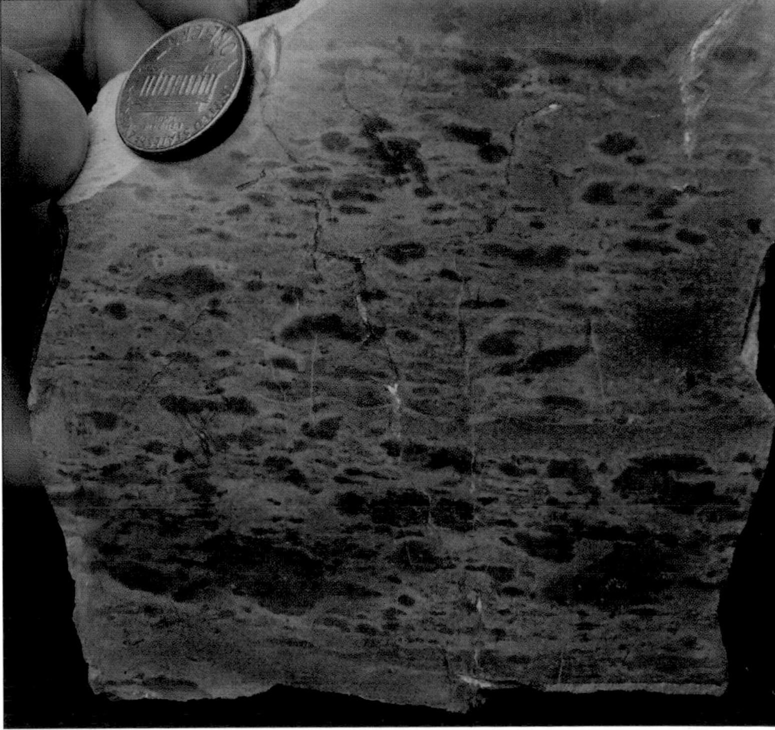

Ozarkcollenia replaced with epidote. Slice through a chunk from Reynolds County, Missouri. (Value range F).

Ozarkcollenia with its calcite replaced with epidote, Reynolds County, Missouri. Epidote usually is not associated with fossils; however, it is a mineral that may owe its existence indirectly to biogenic phenomena more often than realized.

Quartz and Quartz Veins

Because material of the earth has been so highly differentiated compared to meteorites and Moon rocks, the mineral quartz has been extensively concentrated in the earth's crust as a consequence of this differentiation. This concentration has been going on since the Hadean Era and (almost certainly) is related to the fact that water (the hydro of hydrothermal) is such a prevalent material with Earth compared to meteorites, Mars, and the Moon. Precambrian rocks often are shot full of veins of quartz that occupy what appear originally to have been cracks. Such "cracks," now filled with quartz, formed from what is known as hydrothermal activity. This is a process where water-rich, hot gases or fluids containing silicon dioxide from masses of cooling granitic magma deposits quartz. Quartz deposited in cracks as the fluids move upward toward the earth's surface.

Trilobite pygidia ("tail"). Fossils are not common in high grade metamorphic rocks as the process of being drug deep into the earth's crust usually wipes them out. Here is a Cambrian trilobite pygidia in phyllite preserved with epidote. Metamorphism has replaced the calcitic trilobite carapace with epidote in a manner similar to the formation of the above epidote stromatolites. (Value range F).

Epidote with quartz, China. An attractive specimen of these two minerals. During the past decade, excellent mineral specimens like this have become available at reasonable prices from China. (Value range F).

Epidote crystal. This unusually large epidote crystal from North Carolina came from a quartz vein associated with high grade metamorphic rock (schist). It may owe its calcium content to the calcium that was in the parent sedimentary rock—calcite that originally may have come from a biogenic source.

Ruby in zoisite (ruby zoisite). Ruby is a form of aluminum oxide, zoisite is a chromium containing mineral related to epidote. This attractive (and pricey) polished specimen is from an occurrence in Archean rocks that occurs in Tanzania, eastern Africa. (Mundarara Mine, Longido, Kilimanjaro Region, Tanzania. (Value range D).

Back side of same polished sphere.

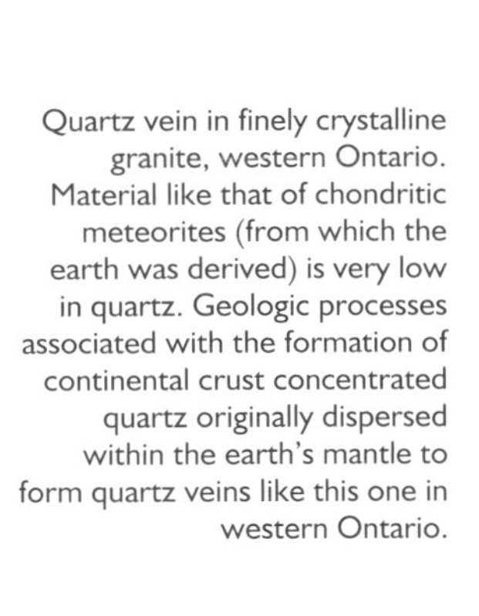

Quartz vein in finely crystalline granite, western Ontario. Material like that of chondritic meteorites (from which the earth was derived) is very low in quartz. Geologic processes associated with the formation of continental crust concentrated quartz originally dispersed within the earth's mantle to form quartz veins like this one in western Ontario.

Ruby in fuchsite. This sphere appears similar to the previous specimen but the green matrix is fuchsite, a chromium containing mica. This attractive group of minerals occurs in Archean rocks of India and is carved into a variety of items like bowls and pendants. This occurrence is more frequently seen than is the more pricey Tanzanian ruby in zoisite. (Value range F).

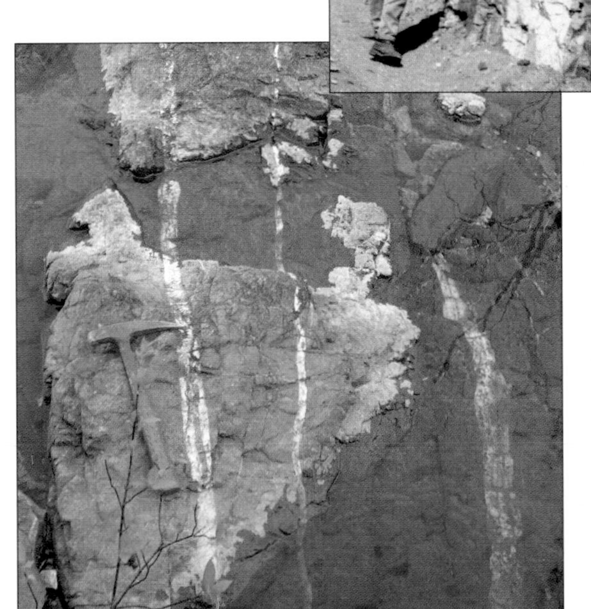

Small quartz veins, east side of Shut-in Creek, Bell Mountain, southern Missouri.

Continuation of one of the same veins across Shut-in Creek.

Sulfide Minerals in Quartz Veins

Quartz veins are especially abundant in Precambrian rocks, where they may be associated with sulfide minerals that can form valuable ore bodies. A number of metallic elements such as silver, lead, zinc, cobalt, copper, nickel, and iron can occur combined with sulfur. Elements like these are referred to as sulfophile elements because they are attracted to sulfur. Such metallic concentrations in quartz veins are known as hydrothermal sulfide deposits and form the target of focused mineral prospecting in many of the earth's ancient rocks.

Entrance to an old silver mine at the west side of Shut-in Creek. Quartz veins like this were followed by mineral prospectors to locate places where valuable metals, occurring as sulfides, were emplaced in the veins. This adit was driven to prospect for silver associated with the quartz veins of previous photos.

Quartz veins in bed of Shut-in Creek, Bell Mountain, southern Missouri.

Enlarged quartz vein near Silver Mine, Missouri. Quartz veins and masses associated with granite offer a clue to the possible presence of valuable elements like silver and gold.

Silver ore. Quartz containing argentiferous (silver bearing) galena, which came from the above prospect and was mined on a small scale in the late nineteenth century.

Ancient rocks of all kinds usually have been preserved from erosion by being drug deep into the crust and thus preserved from erosion and removal. At depths of three or more miles below the surface, the crust is subjected to considerable amounts of heat so that a "cooking effect" takes place, one of the phenomena associated with metamorphism. This burial within the crust will also introduce hot-water-laden vapors—these can contain elements like copper, lead, and silver, sulfophile elements which are attracted to sulfur. Other elements, like tungsten, may also be introduced through hydrothermal activity. Metamorphism associated with deep burial may also "sweat out" and concentrate elements that already were dispersed in the rocks in minute amounts. This appears to be how gold has been concentrated in masses of rock to become mineable. Weathering of a gold bearing rock can concentrate the gold further to form a placer deposit.

Ancient rocks also exhibit more quartz veins and valuable metal mineralization for two reasons. The **first reason** is that there was **more intense thermal activity during the early earth**. There were greater amounts of radioactive isotopes and elements producing more heat at that time. **Second**, ancient rocks often were driven deeper into the crust than were younger ones because the crust was more mobile during the time of the early earth—this could then "sweat out" and concentrate rarer elements like gold.

Hubnernite (black). A manganese tungstate mineral was the ore sought when the Silver Mine's property was mined again in the 1920s and '50s. It is the black mineral associate with quartz. (Value range F).

Tailings at the Einstein Silver Mine, Madison Co., Missouri. This mine worked sulfide-mineral-bearing quartz veins for argentiferous (silver bearing) galena in the late nineteenth century. Later it was worked for hubnernite, a tungsten mineral. Reclamation requirements during the past two decades have mandated that old mining dumps like this be buried and the land contoured as close as possible to its original configuration. Because of this, many interesting and geologically significant areas have recently been covered and made inaccessible. Reclamation of mined areas is often necessary and desirable, however it would be nice if consideration of an area's geological significance was incorporated in the reclamation process.

Green Fluorite, Einstein Silver Mine, Madison Co., Missouri. Masses of this mineral, both purple (the usual color) and green occur in the quartz veins of the Einstein Silver Mine. (Value range H).

Green Fluorite, Einstein Silver Mine. (Value range H).

Quartz masses in Archean rocks, NWT Canada. Metallic sulfide occurrences in quartz are associated with this locality.

Quartz vein fragments carrying chunks of bornite and chalcopyrite, Nonacho Lake, NWT Canada.

Chalcopyrite and bornite from metallic sulfide zone in the quartz vein shown above.

Quartz, variety amethyst with a hematite cap, Thunder Bay, Ontario. Large quantities of amethyst originate from quartz veins and masses that occur in Archean granite near Thunder Bay, western Ontario. (Value range G).

Smoky Quartz, China. Usually associated with pegmatites, smoky quartz is quartz that has had its crystalline lattice modified by exposure to radioactivity. (Value range F).

Druzy quartz (amethyst), Magaliesberg Mountains, Gauteng Province, South Africa. (Value range F).

Druzy quartz (amethyst), Magaliesberg Mountains, South Africa. This quartz druze is probably **not** of hydrothermal origin.

Glossary

Dolostone or Dolomite. A rock (or mineral) composed of calcium magnesium carbonate usually formed in a shallow water, marine environment. Dolostone, when metamorphosed, forms marble like that formed from limestone.

Placer Deposit. A concentration of heavy (or dense) minerals like gold found in a stream bed. Lighter material has been washed away with the denser material being left behind and concentrated. Placer gold is the gold that fueled the gold rushes of the nineteenth and early twentieth centuries.

Regolith. Loose material on the surface of a planet, usually produced as a consequence of some form of weathering. In reference to spinel crystals that occur in regolith, these formed from the weathering and the solution of spinel bearing marble.

Druzy quartz, Missouri Ozarks. Younger than most of the other quartz shown here, this originates from Cambrian age rocks of the Missouri Ozarks where it can locally be very common. This quartz appears to have no connection with any hydrothermal activity. Its origin is somewhat puzzling. (Value range F for group).

Native gold in quartz. Unlike the previous quartz druze occurrence, this quartz is of hydrothermal origin, the gold having been introduced in some sort of vapor that came from elsewhere within the earth. (Value range F).

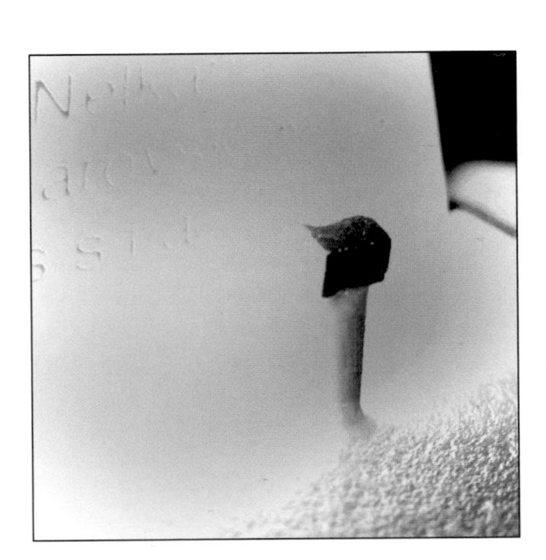

Platinum crystal. Not only gold, but even rarer elements of the platinum group can be associated with rocks of the early earth. These elements may have been "sweated out" from huge volumes of rock and concentrated when the rocks were metamorphosed during deep burial within the earths crust. (Value range E).

Chapter Nine

Xenoliths in (Igneous) Granite

Many granites contain chunks or dark blobs of a different material, which (often) are pieces of rock plucked from the margin of the magma chamber from which the granite originated. **These chunks or blobs are known as xenoliths**. Such occurrences are "foreign rocks" (from *xenos* (g.) = stranger) compared to the granite that contains them. Some granites, however, formed from a process known as granitization, a metamorphic process which, upon deep burial in the crust over long spans of geologic time, converts masses of sedimentary rock like thick shale sequences into granite. These granites, so-called metasomatic granites, usually have a gneiss-like appearance to them and may be more common than is undoubted igneous granite. There is a series of mid-Precambrian granites in the Missouri Ozarks that definitely are not metasomatic, they really did form from the cooling of molten felsic magma within the earth's crust. Such truly igneous granites are the ones with undoubted xenoliths.

In igneous granites are found xenoliths that appear to be pieces of rock (often mafic) plucked from the magma chamber and incorporated into the granite when it was still in a liquid state. Some of these may be of considerable age; however, their actual age is often debatable because heating and thermal metamorphism has reset the geochons to where they may now give to the age of the granite in which they are embedded. At one time, some xenoliths were candidates for being some of the original crust of the earth and they are still being looked at closely in this regard. There is an international competition underway to discover the oldest sample of the earth's crust. Excitement over when and where this will happen has led to a scientific quest.

Olivine xenoliths in basalt. These (geologically young) xenoliths are composed of the green mineral olivine and occur in basalt flows from southern Arizona. The olivine masses are believed to have been "plucked" from the earth's upper mantle and carried upward with the basalt magma, where it reached the surface. (Value range F).

Xenoliths in granite. Archean granite (Keewatin age, 3.2 billion years) containing two xenoliths of metasedimentary rock (one to the left of the man and one to the right of his head). Notice the layering in the xenolith at the left, perhaps it originally was a sedimentary rock. Xenoliths in very old granite like this have been candidates for the world's oldest rocks and even samples of the earth's original crust. These are xenoliths in a true igneous granite that long ago cooled from a felsic magma.

Mafic xenoliths in pink granite, central Manitoba. The black rock is mafic rock of an Archean greenstone belt. Granite has intruded the mafic rock (which originally may have been either basalt or greywacke) and chunks of it have become embedded in the granite. As is typical of many of these ancient xenoliths, the boundary between the granite and the xenolith is sharp and well defined rather than being blurry as might be expected from partial melting. Note the mafic fragment that has separated from the larger mass. The granite is about three billion years old. The mafic rock obviously is older.

Xenoliths in metamorphosed granite gneiss(?) *or are they* **deformed clasts in what originally was a sedimentary rock** now converted to gneiss. Severe metamorphism of very old rocks often makes it difficult to determine just what really happened and what the rocks really represent. These two interpretations of this Archean outcrop in western Ontario are equally valid.

Xenoliths in Archean granite gneiss, **or** is this a granitized breccia? Both interpretations again are equally valid. Western Ontario.

Rectangular granite xenolith in Archean (Keewatin) granite, northern Minnesota.

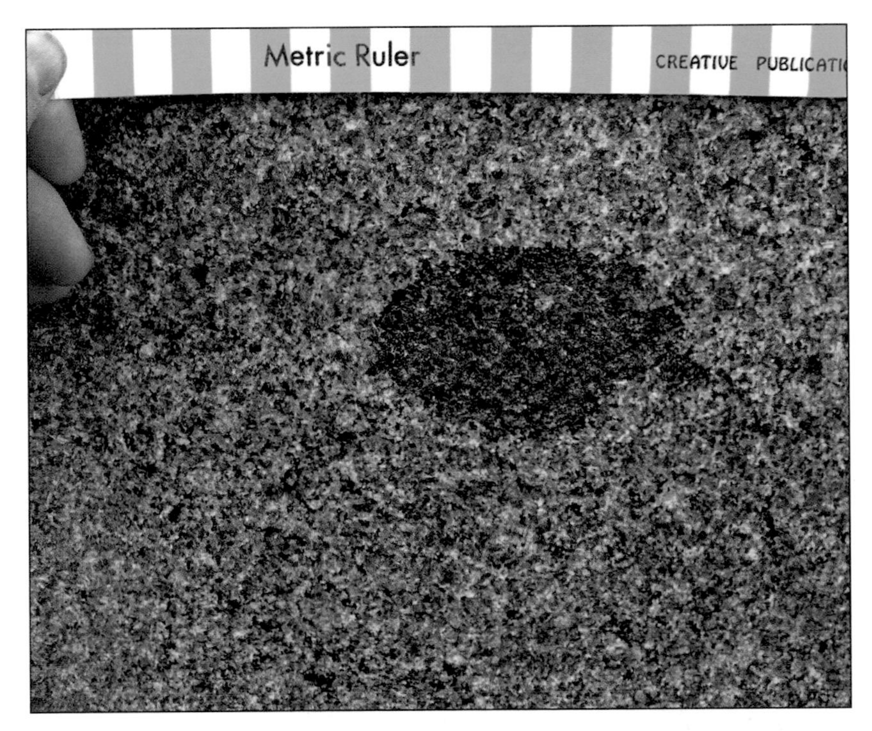

Mafic xenolith? In granite? Or this may be a concentration or clumping of mafic minerals in the felsic magma from which the granite formed. The granite is from southwestern Minnesota, at one time the oldest (known) rock in North America and still one of the older ones at 3.5 billion years.

Glossary

Geochron. A date in actual millions (or billions) of years obtained by radioactive age dating.

Granitization. A process whereby rock like shale or siltstone, through deep burial in the earth's crust over long time intervals, is converted to granite or gneiss. Many granites and granite-gniess masses appear to have formed by this process rather than from the cooling of molten rock (magma).

Magma Chamber. A site within the earth's crust or mantle where molten rock (magma) is located. In reference to xenoliths, which originated from the walls of magma chambers and were "plucked" by a magma mass to be incorporated into it.

Metasomatic Granite. Granite that has been formed from some sort of granitization process.

Elephant rocks, Graniteville, Missouri. These boulders eroded from a 1.4 billion year old granite intrusive to form what resemble (to some persons) pink elephants. Relatively young, compared to the previously shown granites, this granite formed from a felsic magma and unlike many granites of Precambrian age are not granitized sediments.

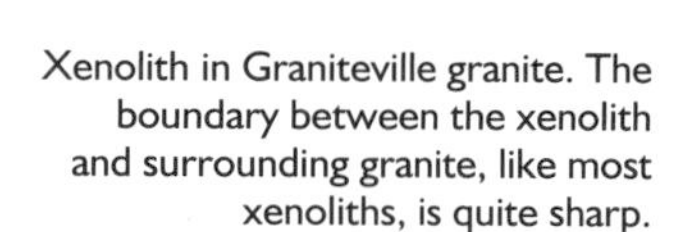

Xenolith in Graniteville granite. The boundary between the xenolith and surrounding granite, like most xenoliths, is quite sharp.